PLASTICITY OF METALS AND ALLOYS

PLASTICITY OF METALS AND ALLOYS

V. V. PUSTOVALOV

Nova Science Publishers, Inc.
New York

Copyright © 2009 by Nova Science Publishers, Inc.

All rights reserved. No part of this book may be reproduced, stored in a retrieval system or transmitted in any form or by any means: electronic, electrostatic, magnetic, tape, mechanical photocopying, recording or otherwise without the written permission of the Publisher.

For permission to use material from this book please contact us:
Telephone 631-231-7269; Fax 631-231-8175
Web Site: http://www.novapublishers.com

NOTICE TO THE READER

The Publisher has taken reasonable care in the preparation of this book, but makes no expressed or implied warranty of any kind and assumes no responsibility for any errors or omissions. No liability is assumed for incidental or consequential damages in connection with or arising out of information contained in this book. The Publisher shall not be liable for any special, consequential, or exemplary damages resulting, in whole or in part, from the readers' use of, or reliance upon, this material.

Independent verification should be sought for any data, advice or recommendations contained in this book. In addition, no responsibility is assumed by the publisher for any injury and/or damage to persons or property arising from any methods, products, instructions, ideas or otherwise contained in this publication.

This publication is designed to provide accurate and authoritative information with regard to the subject matter covered herein. It is sold with the clear understanding that the Publisher is not engaged in rendering legal or any other professional services. If legal or any other expert assistance is required, the services of a competent person should be sought. FROM A DECLARATION OF PARTICIPANTS JOINTLY ADOPTED BY A COMMITTEE OF THE AMERICAN BAR ASSOCIATION AND A COMMITTEE OF PUBLISHERS.

Library of Congress Cataloging-in-Publication Data

ISBN 978-1-60456-965-0

Available upon request

Published by Nova Science Publishers, Inc. New York

Contents

Preface	vii
I. Introduction	1
II. History of the Problem	3
III. Discovery of the Influence of a Superconducting Transition on the Macroscopic Characteristics of Plastic Deformation	7
IV. Basic Experimental Characteristics of the Effect	17
V. Effect of a Superconducting Transition on Deformation Twinning	43
VI. Correlation Between the Characteristics of the Effect and the Superconducting Properties	49
VII. Basic Theoretical Concepts of the Influence of a Superconducting Transition on the Plasticity	69
VIII. Direct Experiments	87
IX. New Method for Studying the Mechanisms of Low-Temperature Plasticity	93
X. Applied Aspects	99
XI. Experiments on HTSC	107
XII. Conclusions	109
References	111
Index	125

PREFACE

The results of observations and investigations of a new phenomenon—changes in the macroscopic characteristics of the plastic deformation of metals and alloys at a superconducting transition—are systematized. In these works it is shown for the first time that the electronic drag of dislocations accompanying low-temperature deformation is effective. The main experimental features of the phenomenon—the dependences of the characteristics of the change in plasticity at a superconducting transition on the stress, deformation, temperature, deformation rate, and concentration of the alloying element in the superconductor—and results indicating a correlation between the characteristics of the effect and the superconducting properties are presented. Experiments clarifying the mechanisms of the phenomenon are analyzed. A brief exposition of the theoretical investigations of the electronic drag of dislocations in metals in the normal and superconducting states and the influence of a superconducting transition on the plasticity is given. The theoretical results are compared with the experimental results. Examples of the application of the effect as a new method for investigating the physical mechanisms of low-temperature plastic deformation are presented. The applied aspects of the phenomenon are discussed separately.

In memory of Prof. V.I Startsev (November 27, 1913-December 26, 1988) and Prof. Taira Suzuki (January 3, 1918–December 19, 1999), who stood at source of these investigations

I. INTRODUCTION

In 1968 two groups of experimentors at the Institute of Solid-State Physics of Tokyo University [1] and at the Physicotechnical Institute for Low-Temperature Physics in Kharkov [2] independently and virtually simultaneously discovered, using different methods, that the macroscopic characteristics of plastic deformation of metals change at a superconducting transition. These were the first experiments to show the effectiveness of electronic drag of dislocations accompanying low-temperature macroscopic deformation of metals and alloys. This moment marks the start of intense experimental and then theoretical investigations of the effect. Similar observations were made subsequently in other macroscopic experiments—under the conditions of creep [3,4] and stress relaxation [5–7]. In the time since the first experiments on the observation of the effect of a superconducting transition on plasticity were performed, extensive investigations of this effect have been conducted in various laboratories in different countries. A series of works was conducted in the USSR (in Ukraine and Russia) and Japan, where the investigations were started, and interesting works were performed in the USA, Canada, Austria, Argentina, and France. Subsequently, a large number of works seeking to prove a connection between the effect and the dislocation-electron interaction and experimental investigations of the mechanisms of the change in plasticity at a superconducting transition were performed and theory was constructed. In a number of works the effect was used as a new method for studying the mechanisms of low-temperature plasticity, and the influence of a superconducting transition on deformation hardening, fatigue, friction, and wear was studied. This could be of value for applications. A generalization of some early works is contained in several reviews [8–15], the last of which was published in Japan in 1985 [15]. Recently, a number of interesting and fundamental investigations have been performed. Together with the early works, which were not incorporated in the reviews mentioned above, they strongly broaden our understanding of the influence of the superconducting transition on plasticity. In this connection a detailed exposition, which is complete as possible, of the present status of the question is timely. Such an attempt is made in the present review, focusing on the systematization of the experimental data.

II. HISTORY OF THE PROBLEM

The question of the mechanical effects occuring at a superconducting transition was initially discussed on the basis of a thermodynamic analysis.16 This analysis showed that changes of the volume, thermal expansion coefficient, and elasticity moduli should occur at a transition of a metal into the superconducting state. The calculations showed that these differences should be very small. For this reason, they were not observed for a long time with the existing measurement techniques. Since the characteristics of inelastic plastic deformation depend on the elastic moduli, it seemed that the influence of a superconducting transition on the plasticity should be sought in changes of the elastic moduli. The development of an experimental technique and the use of an ultrasonic method made it possible to measure the elastic moduli and their changes to a high degree of accuracy (10^{-7}) and to determine small changes of the elastic moduli at a superconducting transition [17] In vanadium single crystals the shear modulus $c_{44}=G$ undergoes the largest change at a superconducting transition. The relative change of this modulus is $\Delta c_{44}/c_{44}=10^{-4}$ at $T=1.5$ K. A similar result was obtained for niobium. The greatest change in the quantity $(c_{11}-c_{12})/2$ occurs in lead—3×10^{-5}. Thus the thermodynamically equilibrium changes of the elastic moduli at a superconducting transition are very small. If the macroscopic characteristics of plastic deformation (the yield point and the flow stress) at a NS transition are assumed to change in the same measure as the changes in the moduli (for example, 10^{-5}), then the differences of these characteristics will be much smaller than the sensitivity of modern detecting apparatus.

An analysis of the plasticity from the standpoint of the intensively developing dislocation models was found to be more fruitful and encouraging. Plastic deformation is a complicated process, which includes dislocation motion (thermally activated, quantum, and dynamical) and the appearance, annihilation, and interaction of dislocations. In such an approach it becomes possible to discuss the influence of a superconducting transition on the plasticity, since the dynamical behavior of a dislocation is determined not only by the interaction with phonons but also with the conduction electrons. This was indicated by the first correct theoretical investigations of the electronic drag of dislocations, published in 1966 [18,19]. Reference 18 also contains qualitative considerations concerning the situation in the superconducting state. Experimental works studying ultrasonic absorption due to dislocation processes started to appear at the same time (1965–1966). It was shown that the amplitudeindependent dislocation absorption of ultrasound is due to oscillations of a dislocation segment, and nonlinear effects (amplitude-dependent absorption of ultrasound) are due to the detachment of dislocations from pinning points [19]. These observations stimulated the development of the theory of dislocation-electron interaction. The macroscopic process of plastic deformation, characterized by the yield point and the flow stress, is due to translational motion of dislocations over much larger distances. For example, for stresses below the yield point in copper single crystals the dislocation travel distances at 77.3 K reach 2000 µm [20]. Above the yield point, at stage I hardening, the average travel distance of edge dislocations is 600–700 µm in copper and 1000–5000 µm in zinc. At stage II the average travel distance is 100 µm, exceeding by five orders of magnitude the displacement of dislocations under the action of ultrasonic oscillations. Thus the effects of electronic drag of dislocations which are observed in ultrasonic experiments could not be transferred to microscopic plasticity, where free and above-barrier dislocation motion, the interaction of dislocations with pinning centers, and the detachment and multiplication of dislocations are entangled in a complicated manner. Therefore the question of the influence of a superconducting transition on the macroscopic characteristics of plastic deformation remained open. Moreover, experiments on the deformation of lead at 77.3 and 4.2 K, respectively, in the normal and superconducting states did not show any changes [21].

In 1967 two reports were presented at a conference in Tokyo. These reports initiated the direct experimental study of the influence of a superconducting transition on macroscopic plasticity. Measurements of the velocities of dislocations in copper, zinc, and copper alloys were reported in Ref. 22. Near the yield point at 77.3 K these values were 10^3 cm/s. This indicated dynamical drag of dislocations, the main component of the drag at low temperatures being electronic. Subsequently, it was found that the superconducting transition influences not only the dynamical but also the thermal-activation motion of dislocations. The results of a study of the temperature dependence of the yield point τ_0 of single crystals of highly pure lead were reported in Ref. 23, and in crystals oriented for single glide τ_0 was observed to decrease between 20 and 4.2 K. In a discussion of the report in Ref. 23 it was suggested that this decrease is due to a transition of lead into the superconducting state. Subsequently, it was found that the anomalous temperature dependence of τ_0 is a more general phenomenon, is observed inmost metals and alloys, including nonsuperconductors, and is due to an intensification of dynamical effects with decreasing temperature [24]. None the less, the results obtained in Refs. 22 and 23 stimulated the development of these works, the next step in which was direct experiments.

III. Discovery of the Influence of a Superconducting Transition on the Macroscopic Characteristics of Plastic Deformation

A. Experimental Arrangement

The experiments on the observation of changes in the macroscopic characteristics of plasticity at a superconducting transition consisted in deforming a sample at the same temperature below the superconducting transition temperature T_c but in different electronic states or with a superconducting transition during deformation. In most such experiments the change in the electronic state was produced by switching a magnetic field with intensity above the critical value (H_c) on and off. For this the sample was placed inside a superconducting solenoid, which was mounted in the deforming apparatus (Figure 1). A special feature and drawback of this method are that in certain experimental situations (see Sec. IV D) when the magnetic field is switched off the deformed sample can trap magnetic flux and this must be taken into account in the measurements. A limitation of this method of changing the electronic state is that high magnetic field intensities (above 1 T) cannot be used because of possible striction effects in the loading components of the experimental apparatus. This eliminates from the investigations superconductors with critical fields above 1 T, which are most important from the standpoint of applications.

The electronic state can also be changed by passing a current with density above the critical value (I_c) through a sample. This method, which is used in Ref.

25, was found to be convenient for performing subtle experiments in the intermediate and mixed states of a superconductor. This method is limited because the current densities must be low in order to eliminate undesirable thermal and mechanical effects. As a result, just as in experiments using a magnetic field, hard technically important superconductors cannot be studied in this manner.

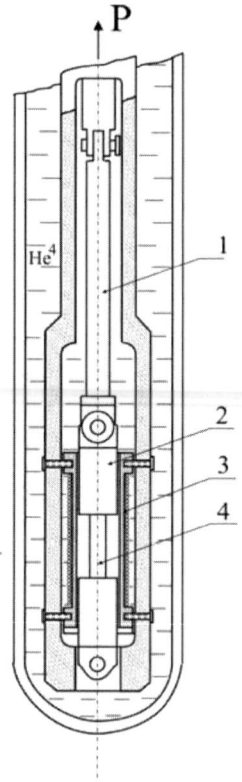

Figure 1. Cryogenic part of the deforming apparatus modified for experiments in a magnetic field: moving rod of the deformation machine *(1)*, sample grip *(2)*, superconducting solenoid secured on a stationary load-bearing tube *(3)*, sample *(4)*.

The transition of a sample from the superconducting into the normal state can be accomplished using a heat pulse, transferring the sample through T_c. This method, which is still unrealized, apparently requires special experimental conditions, because the range of temperature variation must be as small as possible in order to eliminate or reduce to a minimum the possible change of the mechanical characteristics with temperature. True, as will be shown below (Sec.

VI B), the effect weakens as T_c is approached. The difficulties of implementing this method actually offset its potential possibilities for studying superconductors with any values of H_c and I_c. Evidently, this is the reason for the lack of success in observing the influence of a superconducting transition by comparing the mechanical properties of lead [21] at 77.3 and 4.2 K ($T_c \approx 7.2$ K), the stretching curves of niobium [26] at 17 and 7 K ($T_c \approx 9.25$ K), and the kinetics of the development of glide bands in lead [27] at 10 and 4.2 K ($T_c \approx 7.2$ K).

B. CHANGES OF THE CHARACTERISTICS OF PLASTICITY AT A SUPERCONDUCTING TRANSITION

1. Deformation at a Constant Strain Rate

Difference of the Yield Point in the Normal and Superconducting States

The experiments performed to determine the influence of a superconducting transition on the value of the critical resolved shear stress (CRSS) τ_0 of single crystals consisted in applying paired loads, following one after another (with intermediate annealing at $0.8T_m$), on the same sample at T below the superconducting transition temperature T_c up to the CRSS without a field (superconducting state S) and in a magnetic field with intensity above H_c (normal state N). To increase the reliability of the results obtained the sequence of changes of state was changed from one experiment to another and from sample to sample. The first experiments [2,28], performed on 99.9992% pure lead single crystals with stretching axis favorable for easy glide (near [110]), showed (Figure 2a) that the CRSS in the N state is always higher than the CRSS of the same sample in the S state. The average excess over all experiments was about ~30%. In absolute magnitude the difference in the CRSS in the N and S states ($\Delta\tau_{0NS}$) at 4.2 K was 25–710 kPa, the average being 590 kPa. For lead single crystals of the same purity with the stretching axis oriented near [111] these differences are 4.5–39% [29–31].

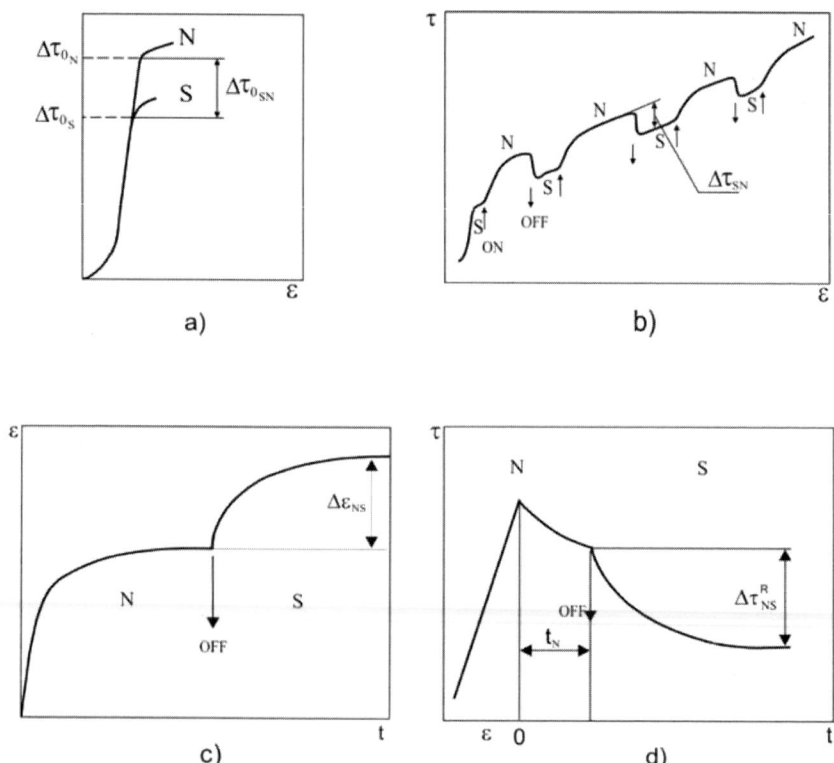

Figure 2. Various manifestations of the influence of a superconducting transition on the macroscopic characteristics of plastic deformation. a, b—under conditions of deformation at a constant rate; on the yield point (a), on the hardening curve (b); under creep conditions (c); under stress-relaxation conditions (d)

The measurements described above are very arduous, which makes it impossible to use them to study different characteristics. At the same time there are definite advantages to the method of repeatedly determining the yield point with intermediate annealing of the sample. A superconducting transition influences the yield point the most. In addition, the experimental arrangement with intermediate annealing makes it possible to remove the possible trapped magnetic flux at *NS* transitions. True, repeated manipulations with a sample made of easily deformable materials, such as lead, indium, and thallium, results in a substantial variance in the correspondingly, $\Delta\tau_{0NS}$, characterizing the influence of the superconducting transition. To eliminate this experiments were performed on a large number of identical samples fabricated from the same indium single crystal [32]. Some samples were loaded in the N state and some in the S state. The elastic limit τ_{00} and the yield point (CRSS) τ_0 were determined from the compression curves. The value of τ_{00} was 46% lower in the S state than in the N state. The value of τ_0 decreased by 41%. Among all superconducting materials studied, the influence of a superconducting transition was greatest in indium.

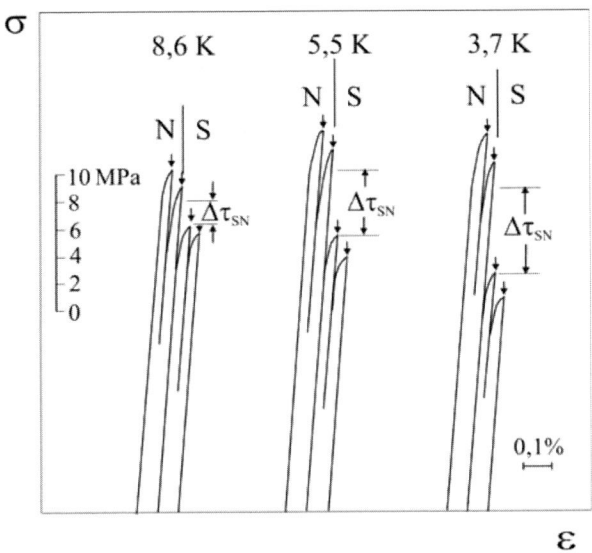

Figure 3. Experimental scheme for repeated determination of the yield point in the normal and superconducting states of high-purity niobium single crystals. The arrows mark the times when the sample was unloaded. Deformation rate 1.5×10^{-4} s^{-1} [33]

Another modification of the experiments near the yield point consisted in repeatedly loading a sample up to stresses slightly above the CRSS (τ_0) without intermediate annealings but with unloadings without a change of temperature [33]. The small decrease of τ_0 with repeated loading was taken into account. Such experiments were performed on highly pure niobium at different temperatures (Figure 3). Taking account of the corrections, $\Delta\tau_{0NS}$ was 0.5 MPa at 8.6 K, which was 2.5% of τ_{0N}. The ratio $\Delta\tau_{0NS}/\Delta\tau_{0N}$ was ~7% at 5.5 K and ~10% at 3.7 K. The same method was used to determine the influence of a superconducting transition on the yield point σ_0 of polycrystals, since heating and annealing can cause recrystallization to occur, changing the results. The measurements on lead polycrystals showed that σ_0 is 1–30% lower in the superconducting state than in the normal state [30].

Change of the Flow Stress at a Superconducting Transition

In the experiments described above, the result of a change of the electronic state was observed but it was impossible to follow the change of the flow stress, clearly illustrating the influence of a superconducting transition on the characteristics of plastic deformation. From this standpoint the method proposed in Ref. 1 has been very fruitful. In the process of continuous deformation, a sample inside a superconducting solenoid was repeatedly transferred from the S into the N state and vice versa (Figure 2b). As one can see, the deforming stress of a single crystal or polycrystal $\tau(\sigma)$ decreases at a NS transition and increases at a SN transition. Qualitatively, such a change is observed at superconducting transitions right up to failure. More than 100 NS and SN transitions can be performed on the hardening curve of sufficiently plastic material, thereby accumulating a definite statistical sample and constructing a curve of the effect versus the deformation, temperature, deformation rate, and other factors. The quantity $\Delta\tau_{SN}(\Delta\sigma_{SN}) = \tau_N(\sigma_N) - \tau_S(\sigma_S)$, where $\tau_N(\sigma_N)$ and $\tau_S(\sigma_S)$ are, respectively, the stationary values of the deforming stress of single crystals (polycrystals) in the normal and superconducting states, is taken as the characteristic of the influence of a superconducting transition on the flow stress. Here are some specific values of the ratio $\Delta\tau_{SN}(\Delta\sigma_{SN})/\tau_N(\sigma_N)$: for lead (polycrystal, 99.9999% purity; 4.2 K)—2.3–3.5% with deformation 0.2–2.7%; indium (polycrystal, 99.999%; 1.8 K)—2.9% with deformation 0.04–0.2%; tin (single crystal, 99.999%; 1.5 K)—0.56% with deformation 2.6%. The drawbacks of the method of repeated superconducting transitions, aside from the abovementioned possibility of magnetic flux trapping after the solenoid is switched off, is that repeated NS and SN transitions influence the strain hardening coefficient (see Sec. X A).

2. Creep

The successful experiments described above [1,2] initiated in 1969 like experiments, performed in Kharkov by two groups of investigators, to observe the influence of a superconducting transition on the creep [3,4]. The first experiments were performed as follows: a stress above the yield point was applied to a sample, which at $T<Tc$ was in a normal state due to a magnetic field, and the curve of nonstationary creep was recorded. When the process became stationary (creep was established), the sample was transferred into the superconducting state. The creep rate increased sharply at this moment (Figure 2c). Such a change of the creep curve at the steady stage occurs once. To repeat the effect the sample must be transferred into the N state (which at the steady stage has virtually no effect on the creep curve) the load must be increased, a new creep curve obtained, the NS transition repeated, and so on. The additional deformation $\Delta\varepsilon_{NS}$ between the stationary stages in the N and S states was chosen as the characteristic of the influence of a NS transition on creep. Lately, a more informative characteristic of the effect has been determined—the increment of deformation $\Delta\varepsilon_{NS}$ over a time interval during which the creep rate $\dot{\varepsilon}_S(t)$ in the superconducting state reaches $\dot{\varepsilon}_N(t)$ at the moment of the NS transition. If creep is produced in one state, then for sufficiently high identical stresses the deformation in the S state is 20% greater than that in the N state. At the moment of the superconducting transition at the steady stage the creep rate increases sharply, the ratio of the creep rates $\dot{\varepsilon}_S$ in the S state to $\dot{\varepsilon}_N$ in the N state is 12–50 in polycrystals of 99.9992% pure lead and 100–300 in 99.9992% lead single crystals [3,34]. At the nonsteady stage, at a NS transition the creep rate also increases and an additional increment to deformation occurs but its magnitude is much smaller than at the steady stage [35]. At the nonsteady stage of creep a reverse SN transition decreases the creep rate, so that a change in the process can be observed repeatedly (Figure 4). The quantities $\Delta\varepsilon_{NS}$ and the ratio $\dot{\varepsilon}_N / \dot{\varepsilon}_S$ with a NS transition depend on the creep rate at the moment of the transition.

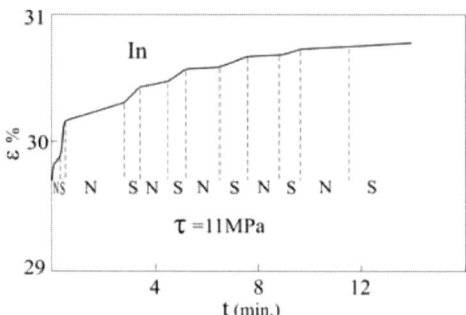

Figure 4. The creep curve of In single crystals under a constant load with repeated NS and SN transitions at the unsteady-state stage [35]

In summary, the most useful information on the characteristics of the effect under creep conditions can be obtained at a *NS* transition at the steady stage. Since this transition is produced, as a rule, by switching a magnetic field off, in cases where magnetic flux is trapped (see Sec. IV D) this method results in substantial errors in the values of $\Delta\varepsilon_{NS}$.

3. Stress Relaxation

The first experiments were performed on two different polycrystals, each of which was deformed in one of the states with periodic stopping and determination of the degree of relaxation [5]. The amount by which $\Delta\sigma$ in the S state exceeded $\Delta\sigma$ in the N state was almost at the limit of accuracy of the experiment.

In 1970 three groups of investigators observed unambiguously and virtually simultaneously the influence of *NS* and *SN* transitions on stress relaxation in experiments performed on one sample [5–7]. Three types of experiments were performed. In one type, [5] during relaxation of the sample in the S state, a magnetic field transferring the sample into the N state was switched on at some time t_S. For small values of t_S this resulted in a sharp decrease of the relaxation rate $\dot{\sigma}$ ($\dot{\sigma}_N < \dot{\sigma}_S$), and for large values of t_S relaxation stopped ($\dot{\sigma}_N = 0$). In another form of the experiment a sample was deformed alternately in the N and S states [38], and the stress was allowed to relax completely in each state. Comparing the degree of relaxation ($\Delta\sigma$ for polycrystals and $\Delta\tau$ for single crystals) in different states showed that for close flow stresses $\Delta\sigma_S$ ($\Delta\tau_S$) is much greater than $\Delta\sigma_N$ ($\Delta\tau_N$). For example, $\Delta\sigma_S/\Delta\sigma_N$ is 1.2–1.5 at 4.2 K in polycrystals of 99.9995% pure lead and 20 in single crystals of 99.9995% pure lead. In the third type of experiment (Figure 2d) the sample was initially relaxed in the N state, and at time t_N the sample was transferred into the S state. In the process the relaxation rate increased sharply, as result of which the degree of relaxation increased. For small t_N the relaxation rate increased by two orders of magnitude, and for large $\Delta\tau_N$ it increased by one order of magnitude. The third type of experiment was found to be most informative, and subsequently it was used to study various characteristics [7,36–39] The quantity $\Delta\sigma_{NS}^R$ ($\Delta\tau_{NS}^R$) was taken as the characteristic of the influence of the superconducting transition on stress relaxation.

The changes of stress relaxation at a NS transition are essentially similar to the changes observed under creep conditions. Just as under creep conditions, with stress relaxation the effect is best observed at a NS transition, so that it is necessary to check for possible magnetic flux trapping. Additional deformation is required for repeated observations. The jump in the relaxation rate and the additional stress relaxation depend strongly on the relaxation time before the NS transition. Consequently, to construct any curves of $\Delta\sigma_{NS}^R$ ($\Delta\tau_{NS}^R$) the relaxation time must be recorded at a NS transition.

IV. BASIC EXPERIMENTAL CHARACTERISTICS OF THE EFFECT

A. GENERAL AND BASIC CHARACTERISTICS OF THE EFFECT

Investigations performed over many years have shown that a change in the macroscopic characteristics of glide is observed in all superconducting materials, which have been studied, with different values of T_c, crystalline structure, and plastic behavior. Among these materials are pure metals: lead [1,2,4–7,12,28–30,34,35,39–82], indium [32,35,41–43,45,46,49,52,55,61,67,76,83–89], niobium [1,90–97] tantalum [98–100], molybdenum [101–104], aluminum [101],[105–110], cadmium [104], thallium [35,43,88], mercury [88], tin [40,43,87,111–116], vanadium [92,117], and zinc [106,118–120]. A large number of investigations have been performed on various binary lead alloys with different concentrations of the second component and state of the impurity in the solution: lead-indium [37,43,45,52,61,68,69,77,110,121–138], lead-tin [39,56,75,128,130,133,134,139–146], lead-bismuth [12,54,69,75,82,124,128,139,147,148], lead-cadmium [36,139–141,144,149], lead-thallium [12,36,54,139,150], lead-nickel [128,142,145,151], lead-silver [152], and lead-antimony [53,143,146,153]. Only several investigations have been performed on niobium-molybdenum alloys [90,91] and aluminum alloys: aluminum-magnesium [107–154], aluminum-copper [154], aluminum-zinc [154] and aluminum-lithium [109,155–157]. The influence of a superconducting transition on the macroscopic characteristics of plasticity have also been observed in superconducting composite materials—copper/niobium [158], copper/niobium—titanium [159] and copper/niobium—zirconium [159].

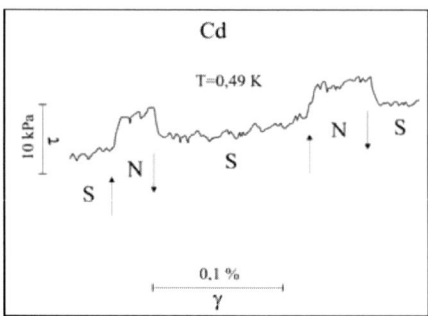

Figure 5. Section of the work hardening curve of a cadmium single crystal (basal glide) with repeated superconducting transitions [104]

A general characteristic is the following: the change in the characteristics of the plasticity at a superconducting transition is pronounced in superconductors manifesting substantial plasticity at low temperatures (lead, aluminum, leadbased alloys, aluminum-based alloys, zinc and cadmium with basal glide). As an example, Figure 5 shows the influence of NS and SN transitions on the deforming stress with basal glide of single crystals of cadmium, which is a superconductor with the lowest, of all materials studied, transition temperature $T_c=0.52$ K. In superconductors where glide is hindered for any reason—pyramical glide in zinc, deformation of tin polycrystals and molybdenum single crystals—the change in the characteristics is weak. Single crystals of highly pure molybdenum can serve as an illustration of such behavion [104]. Another general characteristic is that a change in the deforming stress at NS and SN transitions in the elastic region is not observed with the existing accuracy of stress measurements. Figure 5 shows the results of detailed measurements of $\Delta\tau_{SN}$ at a transition from elastic to plastic deformation. The top panel in Figure 5a shows the initial section of the hardening curve of high-purity aluminum with a change of the electronic state during deformation. The arrows mark the points where a magnetic field $H \geq H_c$ is switched on (↑) and off (↓). It is evident that in a very narrow range of initial deformations (~0.5%) $\Delta\tau_{SN}$ increases rapidly from zero (within the limits of error) in the elastic region up to a constant value. Figure 5b shows $\Delta\tau_{SN}$ versus the applied stress for the same sample. The arrow indicates the stress obtained by extrapolating the hardening curve at the easy glide stage to zero deformation, which corresponds to the yield point. The main growth of $\Delta\tau_{SN}$ refers to a transition from elastic deformation, when microplastic deformation predominates, to macroplasticity, which occurs at a rate set by the deforming apparatus.

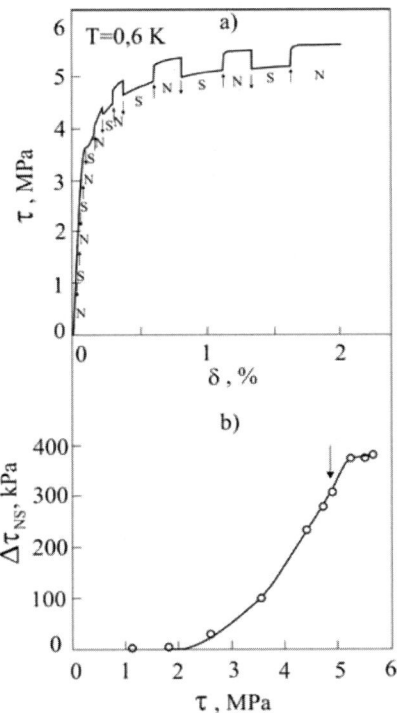

Figure 6. Initial section of the tensile curve of a 99.999% pure aluminum single crystal with repeated *NS* and *SN* transitions (a); $\Delta\tau_{SN}$ versus τ, T =0.6 K; (b) $\dot{\varepsilon}$=1.1×10^{-5} s^{-1} [101]

B. TRANSIENT PROCESSES

1. Deformations with Constant Strain Rate

Recording the curve of work hardening with superconducting transitions during deformation shows that the form of a single jump of the flow stress (processes at *NS* and *SN* transitions) also depends on the plasticity of the material. In plastic superconductors a quite sharp decrease of the flow stress is observed at a *NS* transition (Figures 7a and 7c show examples of a change in τ of this type for indium and aluminum single crystals). The times of the superconducting transition and changes in τ are virtually identical. The flow stress increases at a SN transition. Special experiments performed on lead single crystals at 4.2 K have shown [44] that the time of a change in the electronic state of the sample, determined according to the penetration of the magnetic field into the sample, is less than 1 s. The time $\Delta\tau_{SN}$ of the change in the flow stress was more than 12 s. In low-plastic superconductors (tin single crystals, single crystals of highly pure molybdenum—Figures 7b and 7d) only a decrease of the deformation hardening coefficient θ occurs at a NS transition; no decrease is observed in the flow stress. The effect of further deformation in the S state with a lower value of θ is that $\tau_S(\sigma_S) < \tau_N(\sigma_N)$. At the SN transition θ increases, as a result of which τ increases quite slowly up to τ_N under stationary conditions of deformation.

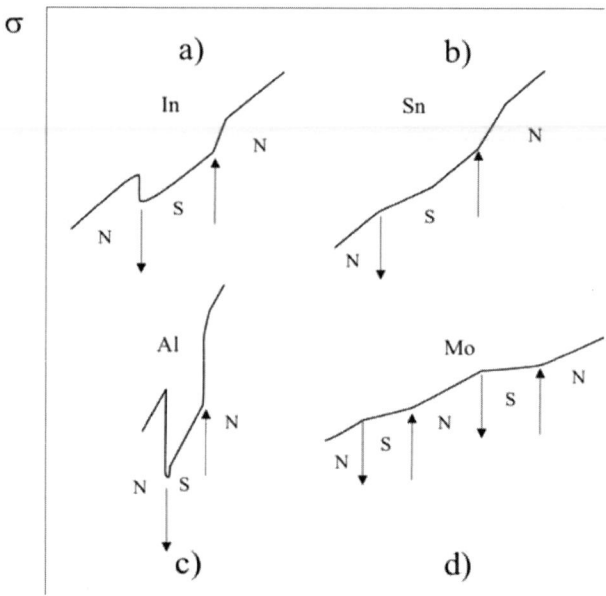

Figure 7. Typical single jumps of the flow stress at NS and SN transitions in different superconductors: indium single crystal (a); aluminum single crystal (b); tin polycrystal (c); molybdenum single crystal (d).

Experiments on 99.9995% pure tin single crystals with stretching axis $\langle 110 \rangle$, which is favorable for glide along the systems $\{100\}\langle 010\rangle$ and $\{121\}\langle 10\bar{1}\rangle$ and is unfavorable for twinning, have shown [52,111] that, in contrast to polycrystals, the plasticity is substantial in the range 3.7–0.5 K. As a consequence, the jump of the deforming stress at a *NS* transition is pronounced, asymmetric, and large in magnitude. Plastic and low-plastic superconductors also differ with respect to the magnitude of $\Delta\tau_{SN}$. Near the yield point $\Delta\tau_{SN}/\tau_N$ can reach several percent in plastic materials, whereas this ratio is a tenth of a percent in low-plastic materials.

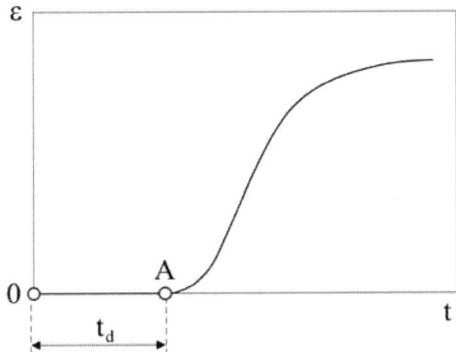

Figure 8. Schematic view of creep strain increment with time during the NS transition, t_d – is the delay time [51]

2. Creep

The study of the kinetics of the change in deformation at a superconducting transition under creep conditions has shown that sharp growth of the deformation occurs with a certain delay time during which the creep rate after the *NS* transition remains unchanged [4,51]. Special experiments have shown that the delay time t_d decreases exponentially with increasing additional load on the crystal in the normal state and decreasnig temperature (Figure 8) [51]. Just as for deformation with a constant strain rate, the increment to the creep deformation at a superconducting transition is different in different metals [35]. In metals which are plastic at low temperatures (lead and indium polycrystals and single crystals, thallium polycrystals) the creep jumps at the onset of a *NS* transition at stresses slightly above the yield point. For low-plastic tin single crystals and polycrystals an appreciable influence of a *NS* transition is observed only at stresses which are much higher than the yield point and approach the ultimate strength. The increment to the creep deformation at a *NS* transition in tin single crystals characterized by substantial plasticity is approximately eight times larger than in polycrystals [52]. Subsequently, the kinetics of the additional deformation

$\delta\varepsilon_{NS}(t)$ of single crystals of highly pure β - tin stimulated by a superconducting transition was studied in detail [115]. The experiments were performed at 1.6 K on crystals with orientation for which dislocation motion through a Peierls potential relief determines the creep kinetics. In general, three stages can be distinguished in the curve $\delta\varepsilon_{NS}(t)$: a delay stage I (just as in lead [4,51], a dynamic stage II, and a fluctuation stage III. The duration and differential characteristics of each stage depend on the creep rate before the NS transition and the magnitude of the total plastic deformation. A quantitative analysis has confirmed the result obtained [116]. Experiments near T_c (3.2 K) have revealed only the delay and fluctuation stages.

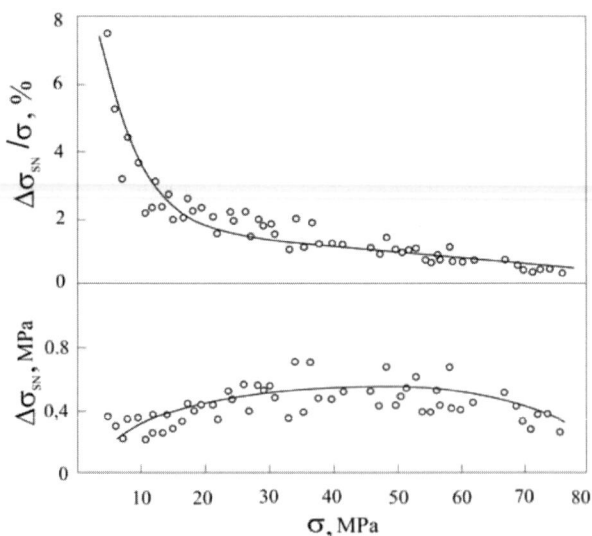

Figure 9. Dependence of $\Delta\sigma_{SN}$ and $\Delta\sigma_{SN}/\sigma_N$ on stress in polycrystalline Pb (99.9995 % pure). T=4.2 K. Deformation rate $7\times10^{-4}s^{-1}$ [29,30]

C. Dependence on Deformation and Stress

1. Jump of the Flow Stress with a Constant Stain Rate

The dependence of the change in the flow stress at NS and SN transitions in a wide range of stresses from the yield point up to the ultimate strength was initially studied on polycrystals and single crystals of highly pure lead [29,30]. These investigations showed that the dependences $\Delta\sigma_{SN}$ ($\Delta\tau_{SN}$) are very different near the yield point and for large deforming stresses (Figure 9).

Growth of $\Delta\sigma_{SN}$ ($\Delta\tau_{SN}$) is observed near the yield point. Subsequent investigations at the initial stages of the deformation of lead have confirmed this dependence [6,40,41,44,45,53,54]. In addition these dependences were found to be insensitive to the deformation rate. Similar behavior was observed for lead single crystals with small additions of tin (1.5×10^{-2}–2.5×10^{-1} at. %) [140], cadmium (1.0×10^{-3}–4.5×10^{-2} at. %) [140], and thallium (3.5×10^{-5}–3.0×10^{-2} at. %) [150]. In addition, the higher the impurity content, the steeper the growth of $\Delta\tau_{SN}$ with increasing deforming stress is. Similar dependences were observed for aluminum single crystals [101] (Figure 6) and in polycrystals of aluminum alloys with magnesium, manganese, and zinc [154].

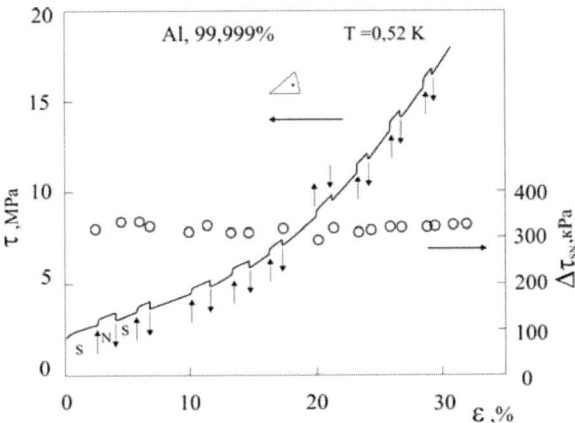

Figure 10. Work hardening curve and $\Delta\tau_{SN}$ versus - deformation dependence taken on Al single crystals. (99.999%).T=0,52 K, $\dot{\varepsilon}$=1,1×10⁻⁵s⁻¹ [101]

As a rule, for high flow stresses $\Delta\sigma_{SN}$ ($\Delta\tau_{SN}$) depends weakly on the stress (deformation), $\Delta\tau_{SN}$ for single crystals is insensitive to the staged nature of the hardening curve (lead, zinc, and aluminum). Published investigations contain two exceptions to this rule. The results obtained for aluminum single crystals [102,107] differ substantially from those presented in Ref. 101. The curve $\Delta\tau_{SN}(\varepsilon)$ was found to be sensitive to the staged nature of the hardening curve.

The decrease of $\Delta\tau_{SN}$ with increasing deformation in Refs. 102 and 107 is probably due to the heating of a quite hard sample during the deformation process as a result of the lower refrigeration power of the ^{3}He cryostat used in Refs. 102 and 107 as compared with Ref. 101.

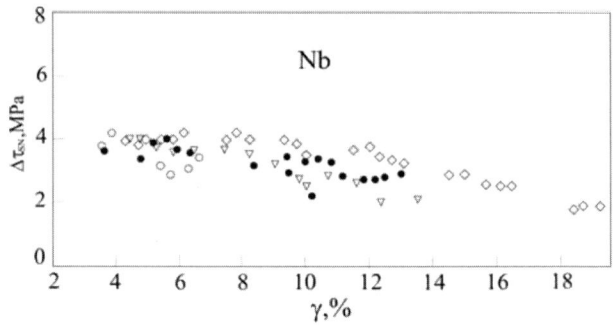

Figure 11. Dependence of $\Delta\tau_{SN}$ on deformation in Nb single crystals. T=4.2 K [90]

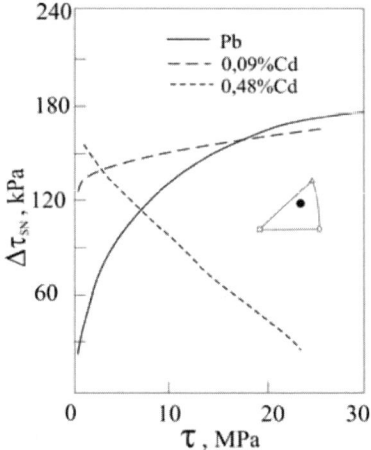

Figure 12. $\Delta\tau_{SN}$ versus stress in Pb and Pb–Cd single crystals, T=4.2 K [90].

The second exception was observed in experiments on superconducting alloys (lead, aluminum, indium, and niobium) and pure niobium (Figure 11). Measurements of $\Delta\tau_{SN}$ ($\Delta\sigma_{SN}$) in these materials have shown an appreciable and in some cases sharp decrease of $\Delta\tau_{SN}$ ($\Delta\sigma_{SN}$) with an increase of the deformation (Figure 12) [12,45,52,54,90,121,122,139,160,161]. Subsequently, it was shown that magnetic flux trapping can occur with repeated superconducting transitions accomplished by switching a magnetic field with intensity above H_c on and off. The presence and magnitude of trapping can depend on the deformation. As a result, the deformation dependence of $\Delta\tau_{SN}$ ($\Delta\sigma_{SN}$) can be distorted (see Sec. IV D for a more detailed discussion).

2. Deformation Jump Under Creep Conditions

Initially, the influence of stress and deformation on $\Delta\varepsilon_{NS}$ was studied in 99.9994% pure lead polycrystals [4] and a sharp growth of $\Delta\varepsilon_{NS}$ with increasing stress was observed above the yield point. A detailed study of $\Delta\varepsilon_{NS}$ versus the applied stress was performed in Ref. 84 on a 99.9999% pure polycrystalline indium. The investigations performed at several temperatures below T_c showed that $\Delta\varepsilon_{NS}$ is very small at stresses below the yield point. As stress increases, $\Delta\varepsilon_{NS}$ also increases, reaching 0.3% at 1.8 K near the ultimate strength. The dependences $\Delta\varepsilon_{NS}(\sigma)$ are qualitatively similar at different temperatures.

A group of superconductors (lead, indium, tin, and thallium) with poly- and single-crystalline structure was investigated in Ref. 35. These experiments showed that the complete curve $\Delta\varepsilon_{NS}(\tau)$ from the start of deformation up to failure is complicated (Figure 13). It is characterized by a definite magnitude of the limiting stress, close to the yield point, where appreciable values of $\Delta\varepsilon_{NS}$ first appear. Then, as the stress increases, three characteristic sections are observed on the curves $\Delta\varepsilon_{NS}(\tau)$. The section I is characterized by sharp growth of $\Delta\varepsilon_{NS}$ in a comparatively narrow interval of stresses. In section II $\Delta\varepsilon_{NS}$ is independent of stress in a wide interval of τ. Finally, section III is once again characterized by a sharp dependence of $\Delta\varepsilon_{NS}$ on τ, though less sharp than on section I. For tin section III was not observed, since the sample failed in section II. Comparing the curves $\Delta\varepsilon_{NS}(\tau)$ with the strain hardening curves for a single crystal showed a definite correlation between these curves. The section I corresponds to the stage of easy glide, and the section II corresponds to the length of the linear hardening stage. One other section of growth of $\Delta\varepsilon_{NS}$ —stage III— corresponds to the softening stage. Therefore the absence of any stage on the hardening curve results in the absence of a corresponding section on the curve $\Delta\varepsilon_{NS}(\varepsilon)$.

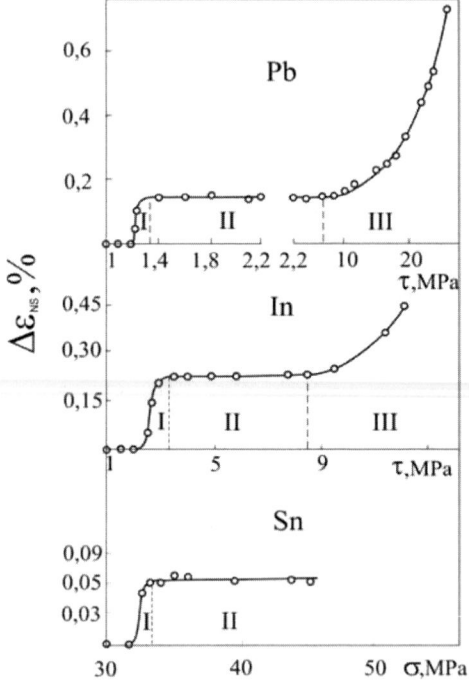

Figure 13. $\Delta\varepsilon_{NS}$ versus stress for lead, indium, and tin [35]

Further investigations of polycrystals of lead with various orientations [57,141] and tin[52] showed that the dependence $\Delta\varepsilon_{NS}(\varepsilon)$ can have a more complicated nonmonotonic character at the easy-glide stage. Experiments on polycrystals of lead alloys with small additions of nonmagnetic Sn (0.4 at. %) and paramagnetic Ni (0.4 at. %) also showed that nonmagnetic and paramagnetic impurities have the opposite effect on the dependence $\Delta\varepsilon_{NS}(\varepsilon)$ [142,151]. In a lead-tin alloy the staged nature of deformation on the curve $\Delta\varepsilon_{NS}(\varepsilon)$ is more pronounced than for pure lead. Conversely, the staged nature is not observed in a lead-nickel alloy. Investigations on lead-tin and -cadmium alloys confirmed the nonmonotonic (bell-shaped) dependence $\Delta\varepsilon_{NS}(\varepsilon)$ at the easyglide stage [141].

In summary, the dependence $\Delta\varepsilon_{NS}(\varepsilon)$ is sensitive to the staged nature of the hardening curve, which is not observed for deformation at a constant rate (4.2.1).

The influence of the total deformation (2–9%) of β-tin single crystals on the creep jump at a NS transition was recently investigated in detail [116]. In contrast to previous works the increment to the deformation over a time interval during which the creep rate $\dot{\varepsilon}_S(t)$ in the superconducting state reaches $\dot{\varepsilon}_N(t)$ at the moment of the NS transition was taken as the integral characteristic of the effect $\Delta\varepsilon_{NS}$. This investigation showed that $\Delta\varepsilon_{NS}$ are monotonically decreasing functions of the deformation ε with maximum values near the yield point and are independent of $\dot{\varepsilon}_N$ in the range $(0.6–1)\times 10^{-5}$ s^{-1}. These functions saturate for $\varepsilon > 6–7\%$.

3. Jump of the Flow Stress Under Stress-Relaxation Conditions

The first investigation of the dependence of the additional relaxation of the stresses $\Delta\tau_{NS}^R$ at a superconducting transition on the applied stress was performed on 99.9995% pure lead single crystals. After the yield point is reached, $\Delta\tau_{NS}^R$ sharply increases and then grows linearly but less rapidly with increasing stress. Similar behavior was obtained for single crystals of lead and lead-indium alloys with several orientations at various deformation rates [38].

D. Magnetic Flux Trapping

Experiments with repeated superconducting NS and SN transitions produced by switching an external magnetic field above H_c on and off during deformation have been widely performed, though before 1981 the appearance of possible trapping of magnetic flux after the magnetic field was witched off was not monitored and not expressly studied. References 79 and 147 are exceptions. In Ref. 79 $\Delta\sigma$ was studied in the intermediate state and trapped flux was observed. This trapping was probably associated with the geometry of the sample. In Ref.

147 the nonmonotonic behavior of the concentration dependence of $\Delta\sigma_{SN}$ was found to be identical to the analogous dependences of the magnetic induction.

The first purposeful investigation of magnetic flux trapping by the structure of a sample when a magnetic field is switched off and of the influence of the trapping on $\Delta\sigma_{SN}$ was undertaken in Refs. 128 and 160. The measurements were performed on a series of alloys—polycrystals of Pb–1.5 at. % Bi, Pb–0.5, 2, and 5 at. % In, Pb–0.4 at. % Ni, and Pb–6 at. % Sn—with stretching along [120]. The extensive set of alloys made it possible to study a sample in a single-phase state with quenching (lead-tin, lead-nickel) and in a two-phase state with aging (lead-tin). The experiment consisted in deforming a crystal, stopping periodically, during which time the magnetization and demagnetization curves were obtained for a sample with increasing and decreasing magnetic field (Figure 14). The curves $B(H)$ were obtained up to fields somewhat above the second critical field H_{c2}. If flux trapping was recorded in the process, then before the next magnetization the sample was heated above T_c ($T\approx 10$ K) and the procedure was repeated but with a different deformation. The change of induction ΔB and the value of $\Delta\sigma_{SN}$ were recorded simultaneously. This made it possible to determine the relative volume of the sample undergoing a superconducting transition and the degree to which $\Delta\sigma_{SN}$ decreases as a result of trapping.

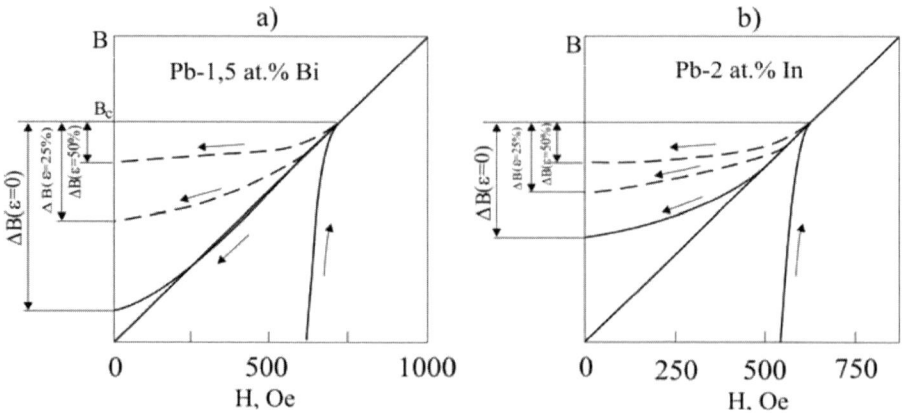

Figure 14. Dependence of magnetic induction B in solid solutions Pb-1.5 at. % Bi (a) and Pb-2 at.% In (b) upon the applied magnetic field under different of deformation [128]

The ratio $\Delta B/B_c$ (B_c is the induction corresponding to H_{c2}, which is independent of the degree of deformation) can serve as a measure of trapping. It was found that the measured deformation dependences of $\Delta\sigma_{SN}$ are completely correlated with the deformation dependences of $\Delta B/B_c$. Figure 15 shows the experimentally measured values of $\Delta\sigma_{SN}$ (filled circles) and the corrected values obtained by dividing $\Delta\sigma_{SN}$ by Bc/B. Such corrected values agree well with the true values of $\Delta\sigma_{SN}$ obtained after the trapped flux is removed by heating. As we can see, the trapped magnetic flux can substantially decrease the measured value of $\Delta\sigma_{SN}$, changing as a result the deformation dependence of $\Delta\sigma_{SN}$. Appreciable flux trapping is also observed in the quenched and aged alloy Pb–6 at. % Sn and in a lead alloy with paramagnetic impurity 0.4 at. % Ni.

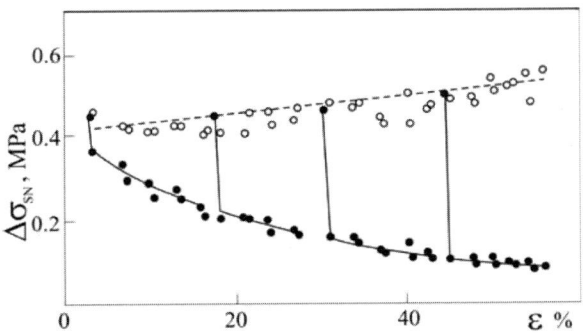

Figure 15. Deformation dependence of the measured (●) and corrected (○) jumps of the flow stress $\Delta\sigma_{SN}$ of the alloy Pb–1.5 at.% Bi. The vertical lines show the first post-heating jump $\Delta\sigma_{SN}$ [128]

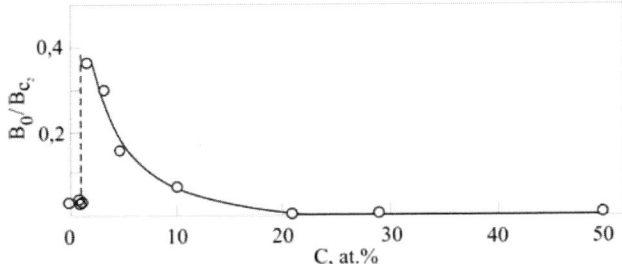

Figure 16. Influence of the indium concentration on the ratio of the trapped flux B_0 to the induction B_c corresponding to the normal state [129]

In summary, the experimental results show that there exists an entire series of situations which are favorable for trapping of magnetic flux in a sample, and the magnitude of the trapped flux can be so large that superconducting transitions occur only in a small volume of the sample. Consequently, the measured change of the plasticity at a superconducting transition in experiments on active deformation and on creep and relaxation of stresses can be substantially understated. In this sense creep and relaxation of stress are more vulnerable, since the experimentally observed effect occurs in them only when the field is switched off. Then, the magnitude of even the first deformation jump (creep) or stress jump (relaxation) is understated in the presence of flux trapping.

For deformation at a constant strain rate the procedure for obtaining the true value of $\Delta\sigma_{SN}$ using heating and measuring the first jump is not always convenient. In the first place repeated heatings substantially prolong an experiment, and in the second place they can introduce changes into the defect structure of the sample. Consequently, the optimum method for monitoring magnetic flux trapping is to measure the curves $B(H)$; the factor $\Delta B/B_c$ is used to make a correlation with the measurements of $\Delta\sigma_{SN}$. It is natural to suppose that magnetic flux trapping can also affect other dependences, which are important for understanding the mechanisms of the change in plasticity at a superconducting transition, for example, on the concentration dependence of $\Delta\sigma_{SN}$. A detailed study of the concentration dependence of $\Delta\sigma_{SN}$ in a wide range of concentrations in the alloys Pb–In neglecting flux trapping revealed a complicated nonmonotonic curve $\Delta\sigma_{SN}(c)$ for deformation at a constant rate [136,138] and $\Delta\varepsilon_{NS}(c)$ under creep conditions [137]. After magnetic flux trapping was found in the deformation experiments, the work in Ref. 129 was performed to obtain the true dependences of $\Delta\sigma_{SN}$ on the deformation and concentration of the alloying element in the alloy Pb–In in a wide range of indium concentrations (up to 50 at. %). For this, the change in the magnetic induction in the sample is monitored simultaneously with direct measurements of $\Delta\sigma_{SN}$. The measurements were performed at 4.2 K on polycrystals of alloys with indium content 0.1, 0.36, 0.54, 1.68, 3.2, 5, 10, 20.7, 27.8, and 50 at. %. At room temperature alloys with these concentrations are solid substitution solutions. Measurements performed on pure lead and alloys with 0.36 and 0.54 at. % In, which are type-I superconductors, have shown that in the undeformed state and with deformation up to 50–60% there is no flux trapping. In type-II superconductors (Pb–1.68, 3.2, and 5 at. % In) strong flux trapping is observed in the undeformed state and intensifies with deformation. Finally, for high indium concentrations (10–50 at. %) trapping of magnetic flux was either absent or very small. The concentration dependence of the parameter $\Delta B/B_c$ (Figure 16), characterizing trapping of the magnetic flux, is nonmonotonic. For In concentrations for which a transition occurs from type-I to type-II superconductivity, sharp growth of flux trapping is observed and the maximum values of $\Delta B/B_c$ are reached. The decrease of $\Delta B/B_c$ for high In concentrations is apparently due to a decrease of the vortex pinning force as a result of an increase of the Ginsburg–Landau parameter at these concentrations [161,162]. The growth of flux trapping with increasing degree of deformation is most likely due to a change of the surface relief (appearance of glide bands and intergrain relief in polycrystals) and the accumulation of defects. Taking account of flux trapping made it possible to construct the true curves of $\Delta\sigma_{SN}$ versus the deformation and concentration, which changed substantially at concentrations where flux trapping is large.

Trapping of magnetic flux was found to be strong in aging lead-antimony alloys [153]. Experiments on the alloys Pb–1.5 at. % Sb and Pb–3 at. % Sb with different preliminary working (quenching, natural and artificial aging) showed that trapped magnetic flux already exists in quenched samples and its magnitude is less than $0.2B_{c2}$. The ratio B_0/B_{c2} increases with the aging time, much and much more sharply for artificial aging than for natural aging. Structural studies and estimates have shown [153] that magnetic flux trapping occurs on precipitates of a second phase, which are effective pinning centers. Just as in the alloy Pb–In, taking account of magnetic flux trapping strongly changed the character of the concentration dependence of $\Delta\sigma_{SN}$ (see Sec. IV F).

The temperature dependences of the characteristics of the change in plasticity at a superconducting transition are also an important characteristic. Consequently, investigations of magnetic flux trapping as a function of the experimental temperature were undertaken. Trapping of magnetic flux B_0 (T) in single- and polycrystals of the alloys Pb–In and Pb–Sn was studied in the temperature range 1.8 K-T_c [134]. In the alloys Pb–1.9 at. % Sn and Pb–6 at. % Sn (with different heat treatment) B_0 increases monotonically with decreasing temperature. Trapped flux arises in Pb-1.8 and 3 at. % Sn single crystals only with large deformations (>50%) and with decreasing temperature (<2.5 K) (Figure 17). In these samples B_0 also depends on the deformation rate. For a low rate ($\dot{\varepsilon}=6\times10^{-6}$s^{-1}) a sharp temperature dependence of B_0 and $\Delta B/B_c$ (where $\Delta B=B_c-B_0$) is observed. It can substantially change the character of the temperature dependence $\Delta\tau_{SN}$ (T), while at the rate 7×10^{-4} s^{-1} such radical changes of $\Delta\tau_{SN}$ (T) are not observed (Figure 18). The differences in the behavior of $B_0(T)$ are explained by the different pinning of magnetic flux by pinning centers. The data obtained on B_0 (T) made it possible to explain by different flux trapping the two types of dependences in $\Delta\sigma_{SN}$ (T) (with and without a maximum), and taking account of flux trapping made it possible to correct the measured values of $\Delta\sigma_{SN}$, as a result of which a single curve $\Delta\sigma_{SN}$ (T) is obtained for all alloys and structures studied (Figure 19).

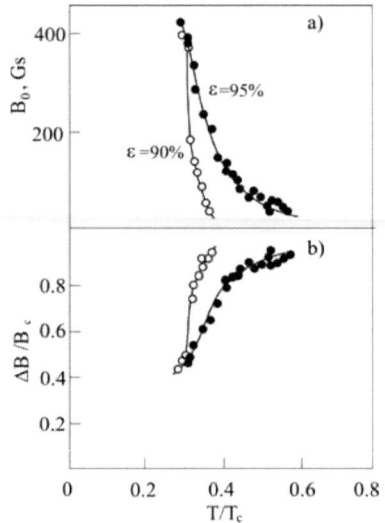

Figure 17. Temperature dependence of the trapped flux B0 (a) and ration of ΔB/Bc in strongly deformed crysrals Pb-3 at.% Sn ; ΔB=Bc-B0 (b)

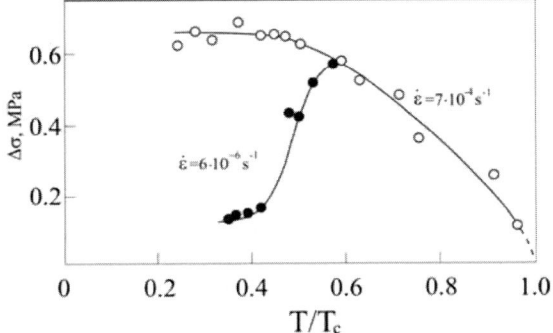

Figure 18. Temperature dependence of the measured values of $\Delta\tau_{SN}$ for Pb–3 at.% Sn single crystals in the case of magnetic flux trapping (low deformation rate) (●) and without flux trapping (high rates) (○), ε=95% [134]

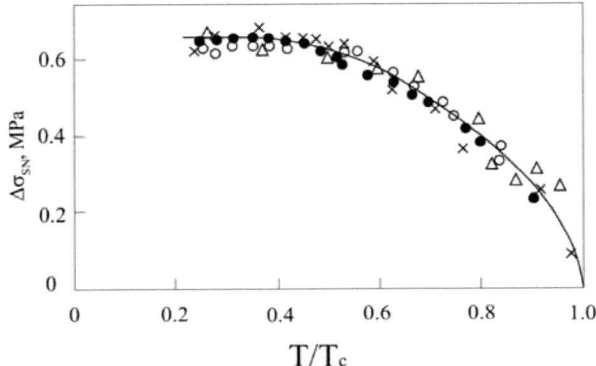

Figure 19. Temperature dependence of the true values of $\Delta\sigma_{SN}$ (T) for Pb–Sn alloys [134]: Pb–6 at.% Sn single crystal, aged for five days at room temperature (●) ; Pb–6 at.% Sn single crystal, aged for one day at room temperature (○); Pb–6 at.% Sn single crystal, quenched (▲); Pb–3 at.% Sn single crystal, annealed (×)

The temperature dependence of flux trapping was studied in Ref. 145 to correct the temperature and deformation dependences of $\Delta\varepsilon_{NS}$ under conditions of creep. Magnetic flux trapping in the strongly deformed alloy Pb–3 at. % Sn substantially changes $\Delta\varepsilon_{NS}$ (T). Interesting observations of the characteristics of magnetic flux trapping in samples with different degrees of deformation, concentration of the alloying element, and structure are contained in a number of works [132,133,149]. As already mentioned, the data obtained on flux trapping made it possible to calculate the true values of $\Delta\sigma_{SN}$ and $\Delta\varepsilon_{NS}$, dividing the measured values of $\Delta\sigma_{SN}$ and $\Delta\varepsilon_{NS}$ by $\Delta B/B_c$. This rescaling is valid, strictly speaking, only for a linear relation between $\Delta\sigma$ and the induction in the sample. The possible nonlinearities which were observed in Refs. 130 and 144 require a more complicated rescaling procedure.

In summary, the method of repeated superconducting transitions is potentially accompanied by magnetic flux trapping, which can distort the values of the parameters ($\Delta\sigma_{SN}$, $\Delta\varepsilon_{NS}$) and the dependences of these parameters on the deformation, concentration of the alloying element, and temperature.

E. STRAIN RATE DEPENDENCES

1. Deformation at a Constant Strain Rate

The first measurements of $\Delta\sigma_{SN}$ with different rates of deformation of lead single crystals were performed in Refs. 6 and 41. The measurements of $\Delta\sigma_{SN}$ were performed with stretching at 4.2 K with rates 8×10^{-5}, 8×10^{-6} and $1,6\times10^{-6}$ s^{-1}. Although the deformation rates were changed in different parts of the deformation curve by more than a factor of 50, no systematic influence of the rate was found. Insensitivity to the deformation rate was also observed in an indium sample. The rate dependences of $\Delta\sigma_{SN}$ for 99.9995% pure lead polycrystals at 4.2 K with stretching were studied in an even larger interval of rates (7×10^{-3}, 7×10^{-4}, 3×10^{-4}, and 2.4×10^{-5} s^{-1}) in Ref. 40, but no rate dependence was found within the limits of accuracy of the measurements. For stretching of niobium single crystals [61] the rate was varied by a factor of 40 ($\sim10^{-4}-10^{-5}$) at 4.2 K and no systematic rate dependence was observed. In a series of experiments on indium [85] the deformation rate was changed by a factor of 200 (2×10^{-3}- 1×10^{-5} s^{-1}), and no differences were found in the values of $\Delta\sigma_{SN}$. The most detailed study of the rate dependence of $\Delta\sigma_{SN}$ was performed on 99.9995% pure lead polycrystals [45]. The rate dependence was determined in the range 2×10^{-3}- 1×10^{-5} s^{-1} – 6.6×10^{-3} s^{-1} at 1.65 K. Two samples were investigated for each rate, and five samples were investigated at the rate 6.6×10^{-4} s^{-1}. The results are presented in

Figure 20. For comparison the same figure shows the dependence $\Delta\sigma_{SN}(\dot{\varepsilon})$ at 4.2 K. This curve was constructed according to Ref. 40. Evidently, in contrast to the results obtained at 4.2 K, where $\Delta\sigma_{SN}$ is virtually rate independent, as temperature decreases to 1.65 K an appreciable dependence of $\Delta\sigma_{SN}$ on $\dot{\varepsilon}$ appears. The values of $\Delta\sigma_{SN}$ increase by a factor of 1.6 when the rate increases by two orders of magnitude.

The measurements of the rate sensitivity of the deforming stress performed separately in the normal and superconducting states are close to the rate dependences of $\Delta\sigma_{SN}$ under study. Such measurements on lead single crystals at 4.2 K were performed in Ref. 47. The deformation rate was changed by a factor of 10 with a base rate of 4.56×10^{-5} s^{-1}. Each point on the curve $\Delta\tau_{10}(\tau)$ is an average of 8–10 measurements. As one can see (Figure 21) the strain rate sensitivity in the N state is higher than that in the S state in a wide range of stresses. A similar result was obtained in Ref. 85 for indium single crystals, in Ref. 63 for lead single crystals at 4.2 K, and in Ref. 163 for pure lead crystals at 2.2 K.

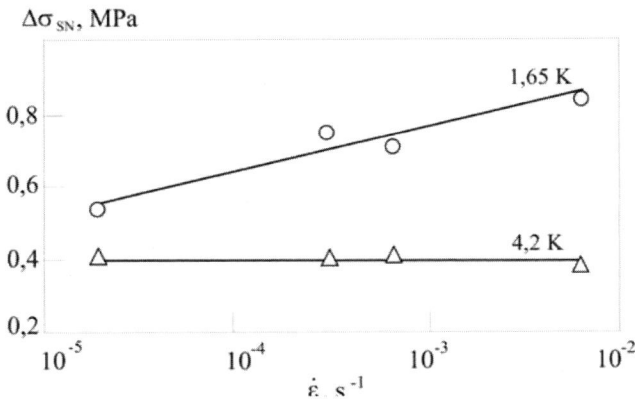

Figure 20. $\Delta\sigma_{SN}$ versus deformation rate for 99.9995% Pb polycrystals at T, K: 4.2 (Δ) and 1.65 (○) [45]

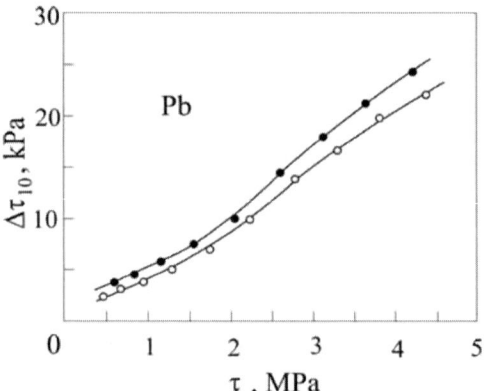

Figure 21. Strain rate sensitivity $\Delta\tau_{10}$ (measurements with a change in deformation rate by a factor of 10) of the flow stress of annealed single crystals of lead as a function of τ in the normal and superconducting states at 4.2 K: N (●), S (○) [47]

The strain rate sensitivity of τ in the normal and superconducting states was studied in Ref. 99 for single crystals of highly pure tantalum. The values of $\Delta\tau_{10}$ were characterized by a large variance, within which the electronic state was not observed to influence $\Delta\tau_{10}$. Among the works studying the rate dependence of τ there is an investigation of single crystals of Al and Al–Mg alloys [107]. The results differ substantially from what was observed for lead and indium. The experiments showed that in a wide temperature range the strain rate sensitivity of $\Delta\tau_{10}$ with a ten-fold change in the strain rate is much greater in the S state than in the normal state. It was noted previously (see Sec. IV A) that in these experiments an anomalous, as compared with Ref. 101, stress and deformation dependences of $\Delta\tau_{SN}$ were observed.

2. Creep

Nonsteady creep under a constant applied stress occurs with continually decreasing deformation rate. The additional deformation $\Delta\varepsilon_{NS}$ at a superconducting transition with a fixed stress depends on the location of the transition on the creep curve and can reach a magnitude equal to the deformation at the transition stage in the normal state, where the creep rate is maximum [27].

As the rate decreases, $\Delta\varepsilon_{NS}$ decreases but does not vanish, even at the steady stage. A detailed investigation of the rate dependence of $\Delta\varepsilon_{NS}$ for polycrystals of highly pure (99.9997%) lead was performed in Ref. 73. For this $\Delta\varepsilon_{NS}$ was determined on different sections of the curve of nonsteady creep which correspond to different instantaneous rates. The stress dependences of $\Delta\varepsilon_{NS}$ were determined for different creep times in the normal state. It was found that in the coordinates $\Delta\varepsilon_{NS}$-$\ln\dot{\varepsilon}$ the curves have the same form irrespective of the magnitude of the stress (Figure 22).

Detailed information on the influence of the starting rate ($\dot{\varepsilon}_N$ at the moment of a NS transition) on the total jump $\Delta\varepsilon_{NS}$ in β-tin single crystals at 1.6 and 3.2 K was obtained in Ref. 116. The main feature of the rate dependence obtained was the presence of two intervals on the rate axis $\dot{\varepsilon}_N$—a comparatively narrow interval of rates of strong and a wide interval of weak rate sensitivity of $\Delta\varepsilon_{NS}$. The influence of the magnitude of the starting creep rate on the value of $\Delta\varepsilon_{NS}$ near T_c remains but is weaker. 1

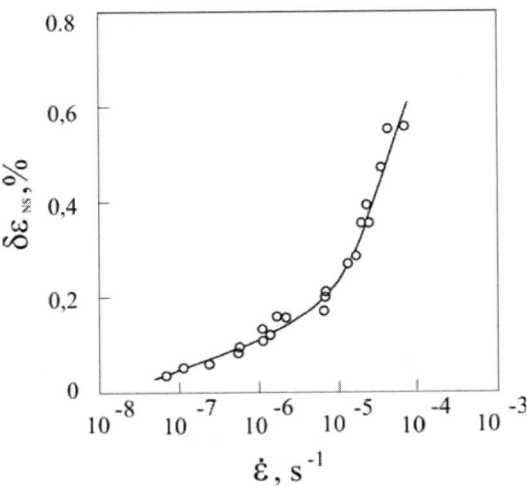

Figure 22. Dependence of $\Delta\varepsilon_{NS}$ on strain rate in Pb polycrystals at 4.2 K [73]

F. CONCENTRATION DEPENDENCES OF THE CHANGE IN THE CHARACTERISTICS OF PLASTICITY AT A SUPERCONDUCTING TRANSITION

Experiments where the influence of the concentration of impurities and alloying elements on the parameters of the changes—$\Delta\tau(\Delta\sigma)_{SN}$; $\Delta\varepsilon_{NS}$ and $\Delta\tau(\Delta\sigma)_{NS}^{R}$.—are very important for understanding the general behavior and especially for determining the mechanism of the influence of a superconducting transition on plasticity. Consequently, after the effect was observed investigations on impurity crystals and alloys appeared. The first experiments were performed on single crystals of the alloy Nb–Mo [90,91] and single crystals of lead with different degrees of purity [6]. It was found that $\Delta\tau_{SN}$ increases as the Mo concentration increases up to 2.5 at. %91 and as the purity of the lead decreases. The sensitivity of $\Delta\tau_{SN}$ to the purity and alloying subsequently initiated a large number of investigations on various alloys in a quite wide range of concentrations. The largest number of investigations was performed on lead alloys possessing substantial plasticity at low temperatures. Lead alloys with thallium (0.205, 0.324, 1.02, 1.4, 3.98, 4.3 at. %), cadmium (0.087, 0.46, 0.475, 0.845 at. %), tin (0.96, 1.93 at. %), and bismuth (0.085, 0.344, 1.8, 3.8, 3.92 at. %) were studied in Ref. 139. The results obtained for all alloys fall on a single curve $\Delta\tau_{SN}/100\,|\delta|$ versus logc, where $\delta = 1/a(da/dc)$ is the dimensionalmismatch, a is the lattice parameter, and c is the concentration. Experiments on lead single crystals with thallium (0.04, 0.4 at. %) and bismuth (0.04, 0.4 at. %) performed at different temperatures also showed that $\Delta\tau_{SN}$ in alloys with 0.4 at. % of the alloying element is larger than in alloys with 0.04 at. % [54]. Polycrystals and single crystals of lead-indium alloys were studied in Ref. 136 in a wider range of concentrations—0.75–0.93, 10, and 12.2 at. %. In contrast to pure metals, $\Delta\sigma_{SN}$ in polycrystals of alloys was observed to decrease with increasing σ, practically in the entire interval of values of σ. This decrease was especially sharp in the alloy with 0.85 at. % indium. The character of $\Delta\sigma_{SN}(c)$ was found to depend on the degree of deformation. For small deformations (low stresses) $\Delta\sigma_{SN}$ was observed to increase with the indium concentration. As the deformation increases, the concentration dependence of $\Delta\sigma_{SN}$ becomes nonmonotonic and has a minimum near 1 at. %. A more detailed investigation of lead-indium alloys up to a concentration 21 at. % indium confirmed the behavior obtained [138]. An even more complicated nonmonotonic concentration dependence of $\Delta\sigma_{SN}$ was observed and investigated in polycrystals of lead-bismuth alloys [147]. The character of the dependence $\Delta\sigma_{SN}$ (c) for different degrees of deformation was the same and exhibited the following characteristic features. For low concentrations (up to 0.73 at. %) $\Delta\sigma_{SN}$ decreases with increasing concentration, reaching minimum values. The maximum value of $\Delta\sigma_{SN}$ is observed at 1.18 at. % and a second minimum of $\Delta\sigma_{SN}$ is observed at 1.5 at. %. At higher concentrations $\Delta\sigma_{SN}$ increases slowly and a diffuse maximum is observed near 10 at. %.

Studies were also performed under conditions of creep [137,164] and stress relaxation [36]. In lead-antimony alloys (0.85, 1.45, and 3 at. %) the samples were studied in quenched and aged states, and the concentration dependences were found to be different in different structural states [164]. A complicated nonmonotonic dependence $\Delta\varepsilon_{NS}$ (c) with a minimum near 2 at. % was observed in lead-indium alloys [137]. This dependence was insensitive to the degree of deformation. Conversely, in investigations performed in the presence stress relaxation [33] for the system Pb–In monotonic growth of $\Delta\sigma_{NS}^{R}$ was observed with increasing concentration up to 10 at. % indium. The discovery of magnetic flux trapping required that trapping be taken into account and the corresponding correction be made in the values of $\Delta\sigma_{SN}(\Delta\tau_{SN})$, which substantially changed (qualitatively and quantitatively) most of the results presented. This made it necessary to perform new investigations to determine the true concentration dependences. Experiments on lead-antimony alloys (0.4, 0.7, 1.5, 3.0, 5.8 at. % Sb) in quenched and aged states were fundamental to determining the true concentration dependences of $\Delta\tau_{SN}$ [153]. Since trapped magnetic flux existed in quenched and especially in aged samples, the measurements of $\Delta\tau_{SN}$ were performed when the magnetic field was switched on for the first time. It was established that in the alloys studied $\Delta\tau_{SN}$ is independent of the aging time, and the concentration dependences of the true values of $\Delta\tau_{SN}$ in the quenched and aged samples are the same and are monotonic, increasing nonlinear functions of the antimony concentration. This result is different from that obtained in Ref. 164.

Figure 23 shows the dependences $\Delta\tau_{SN}(c)$ obtained by different authors for lead-based solid solutions. It is evident that all dependences have a similar form in the concentration range presented. A new detailed investigation was also performed on the system lead-indium in a wide concentration range (0.1, 0.36, 0.54, 1.68, 3.2, 5, 10, 20.7, 27.8, 30, and 50 at. %) taking account of flux trapping [129]. The resulting curve of the true values of $\Delta\tau_{SN}$ in the range 0–50 at. % In is presented in Figure 24. The curve has a diffuse maximum near 5 at. % In. The results obtained for lead with very low additions of impurity should be presented to complete the picture. Measurements on lead with additions of tin (ranging from ~3 ppm up to 104 ppm) established that $\Delta\tau_{SN}$ is concentration independent up to 0.05 at. % [56,152]. Detailed investigations of the same type were performed on lead-tin and lead-cadmium systems [140]. The curves $\Delta\tau_{SN}(c)$ for identical applied stresses (Figure 25) were constructed on the basis of measurements of $\Delta\tau_{SN}(\tau)$. It is evident that for the alloys studied there is a certain "critical" impurity concentration c_{cr} below which $\Delta\tau_{SN}$ is concentration independent and above which $\Delta\tau_{SN}$ increases with concentration. As τ increases, in both alloys c_{cr} shifts to high concentrations. Magnetic flux trapping with repeated superconducting transitions is not observed.

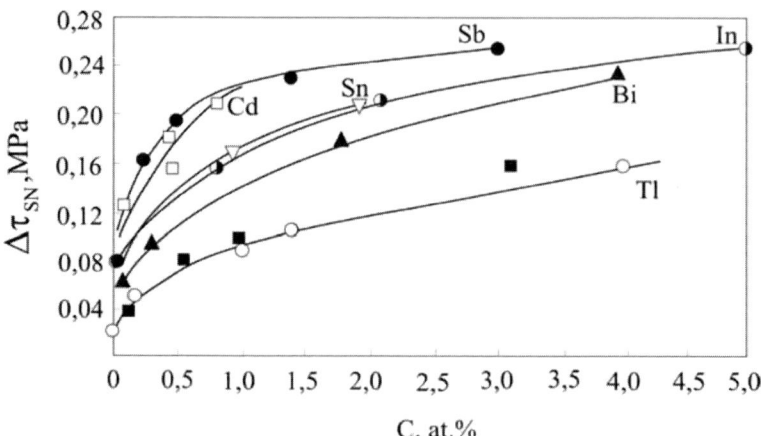

Figure 23. Concentration dependences of $\Delta\tau_{SN}$ in lead-based alloys: Tl (○) [139]; Tl (■) [150]; Bi (▲) [139]; Sn (∇) [139]; Cd (□) [139]; Sb (●) [153]; In (ʋ) [129]

Plasticity of Metals and Alloys 41

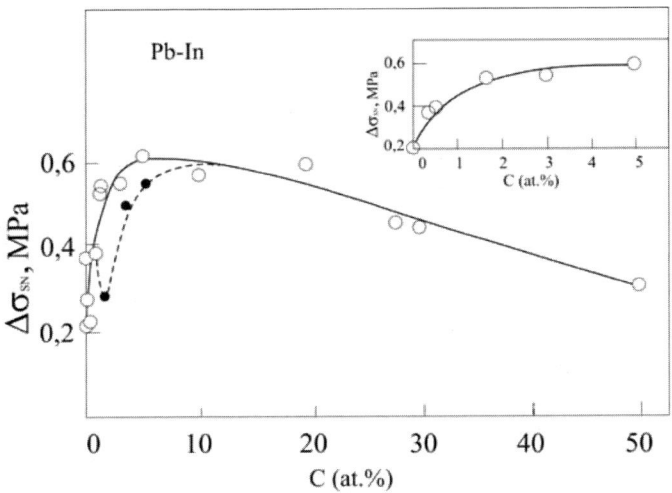

Figure 24. Concentration dependence of $\Delta\sigma_{SN}$ in polycrystals of lead-indium alloys. Deformation 5% (values of $\Delta\sigma_{SN}$ obtained neglecting magnetic flux trapping (●)) [129]

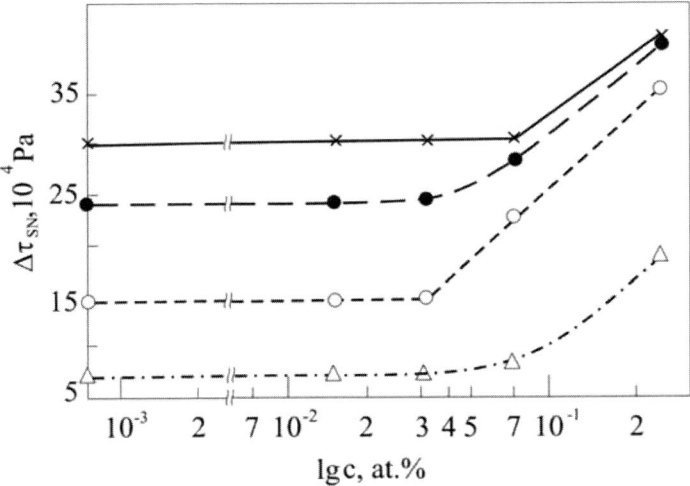

Figure 25. $\Delta\tau_{SN}$ versus tin content in single crystals of lead at low concentrations with different applied stresses: $\tau=0.2\times10^7$ Pa (8); $\tau=0.4\times10^7$ Pa (○); $\tau=0.5\times10^7$ Pa (●); $\tau=1.0\times10^7$ Pa (□) [140].

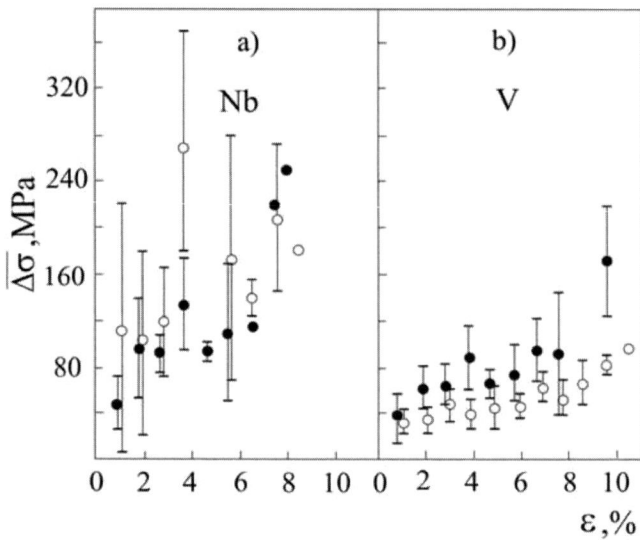

Figure 26. Dependence of the average stress jump upon deformation of Nb single crystals (a); V single crystals (b); superconducting state (●); normal state (○); T=4.2 K, $\dot{\varepsilon}=7{,}3\times10^{-5}$ s^{-1} [92].

Summarizing the results of measurements of $\Delta\tau_{SN}(c)$ we call attention to the presence of three concentration intervals differing by the character of the changes in the true values of $\Delta\tau_{SN}$. For very low concentrations (up to ~10−2 at. %) $\Delta\tau_{SN}$ is concentration independent and equals $\Delta\tau_{SN}$ for a pure superconductor. Up to approximately 5 at. % of the alloying element $\Delta\tau_{SN}$ ($\Delta\sigma_{SN}$) grows nonlinearly with the concentration. From 5 to 50 at. % (one experiment on the system Pb–In [129]) $\Delta\sigma_{SN}$ decreases slowly with increasing concentration.

V. Effect of a Superconducting Transition on Deformation Twinning

Thus far the influence of a superconducting transition on glide processes was studied in detail. There exists another form of plastic deformation—twinning, which appears quite often together with low-temperature deformation. Experiments under twinning conditions are very important for obtaining a complete picture of the influence of a superconducting transition on plasticity.

The process of deformation twinning is characterized by a number of features which distinguish it from gliding. These include high rates of the process, a small specific shift in fractions of Burgers vector, and localization of the process with expansion of the twinning interlayer. Other features which complicate and sometimes make it impossible to study the influence of a superconducting transition on twinning also occur. The twinning process is often combined with glide and it is difficult to determine which part of the effect is due to glide and which to twinning. In addition, as a rule, large-amplitude jumps of the stress correspond to a twinning process on the macroscopic deformation curve. It is impossible to catch, against the background of these jumps, which often follow one after another continuously, the comparatively small changes of the deforming stress.

These difficulties have been overcome with various degrees of success in several experimental works. Studying creep in indium polycrystals and its change at a superconducting transition, the authors of Ref. 84 called attention to the fact that at room temperature plastic deformation occurs by gliding and twinning, and as temperature decreases, twinning occurs more easily and its contribution to deformation increases. Consequently, it is believed that the creep jump observed at a *NS* transition is associated with additional thickening of the twinning interlayers which is due to unblocking of twinning dislocations.

The work described in Ref. 117, performed on 99.98% pure vanadium single crystals ($R_{293}/R_{4.2(N)} \approx 10^2$) under conditions of stretching with rate 3×10^{-4} s^{-1}, was concerned with determining the quantitative differences between twinning in the normal and superconducting states. At 4.2 K plastic deformation of vanadium single crystals with different orientation started with the formation of twins, so that the difference in the stress at which twinning starts was taken as a characteric of the effect of a superconducting transition on twinning. The stress at the onset of twinning was always higher in the normal state than in the superconducting state. This signified that the initial nucleation of twinning dislocations is sensitive to the electronic state of the sample, being facilitated in the superconducting state. The jumps (frequency and amplitude) on the stretching curves were also analyzed. These jumps were due to the twinning and were obtained completely in the normal and superconducting states. The number of jumps occurring per one percent of deformation with different degrees of deformation does not depend on the electronic state of the sample. On the other hand the amplitude of a jump in the superconducting state is 1.3 times larger than in the normal state. Since the abovementioned difference in the stresses at which twinning starts in the *N* and *S* states attests to facilitation of the work performed by the sources of twinning dislocations at a superconducting transition, the difference in the amplitude of the jumps signifies that electronic drag influences the motion of twinning dislocations.

An attempt to study the influence of the electronic state of niobium single crystals ($R_{300}/R_{20}=200$) on twinning under deformation by compression was made in Ref. 95. Measurements of the stress at the first jump (similarly to Ref. 117) showed that in niobium (in contrast to the the result obtained for vanadium in Ref. 117) this stress is lower in the *N* state than in the *S* state.

To determine the reasons for the differences in the results obtained for vanadium [117] and niobium [95] the same experiments were performed in Ref. 92 on niobium and vanadium single crystals under conditions of compression with rate 7.3×10^{-5} s^{-1} at 4.2 K. Both substances under these conditions deformed by twinning along the system $\langle 111 \rangle (11\bar{2})$, which was manifested in the form of the jumps following one after another. Metallographic observations performed on a heated sample showed that a definite number of twinning interlayers with almost the same thickness $\approx$3–5 μm arose at each jump. The deformation curve was obtained either in the *S* state or in the *N* state (longitudinal field 7 kOe); at least five samples were studied in each state. The amplitude of the jumps ($\Delta\sigma$) was chosen as the twinning parameter. In niobium the magnitude of a jump $\Delta\sigma$ was essentially the same in different states, though $\Delta\sigma$ tends to be large in the N state. In vanadium it is more clearly evident that, in contrast to niobium, $\Delta\sigma$ is larger in the S state than in the N state. Such a difference is attributed to the fact that the superconducting transition in niobium and vanadium results in volume changes $\delta = (V_N - V_S)/V_N$ of opposite sign— $\delta < 0$ in niobium and $\delta > 0$ in vanadium. If this is the case, then the data obtained show that the nucleation of twins occurs more easily in crystals compressed in three dimensions (vanadium in the S state and niobium in the N state) than in an expanded crystal (niobium in the S state and vanadium in the N state). It should be noted that the values obtained for $\Delta\sigma$ (especially for niobium) have a large variance, which could mask the real behavior. In addition, estimates have shown that glide plays a large role in a jump. Therefore $\Delta\sigma$ can include not only twinning but also glide. In this sense the functions $\Delta\sigma(\varepsilon)$ can change substantially.

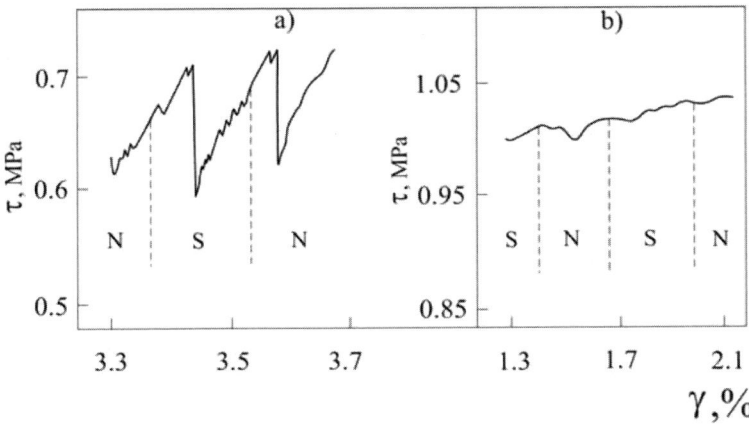

Figure 27. Sections of the compression curves of In–10 wt.% Pb (a) and In–14 wt.% Pb (b) single crystals in the region of pseudotwinning with repeated superconducting transitions. T=1.6 K; $\dot{\varepsilon}$ =2×10−4 s−1 [165]

Subsequent investigations performed on indium and its alloys [32] and niobium [96] did not settle the issue, but they are important for understanding the situation. Single crystals of pure indium (99.999%) and In–Pb alloys (6 and 8 at. % Pb), which under compression in the [001] direction deform solely by twinning up to complete reorientation of the samples, were investigated. The experiments were performed in liquid ^{4}He (1.7 K) and liquid ^{3}He (0.48 K). The influence of the electronic state of the sample on the stress σ_t at which twinning starts was determined. No appreciable influence of the electronic state with a larger or smaller variance in the the values of σ_t was observed in either pure indium, where substantial glide precedes twinning, or in the alloy In–6 at. % Pb, where plastic flow right to deformation ~10% occurs exclusively by twinning. The influence of a superconducting transition on the deforming stress was studied in Ref. 165 in the alloys Pb–6 at. % In and Pb–8 at. % In, which possess superconductivity due to twinning. For the stress measurement accuracy 2.5×10^{-3} MPa no influence of *NS* and *SN* transitions could be observed (Figure 27). Thus, the experiments on In–Pb alloys did not yield obvious and just as convincing as in the case of glide evidence of an influence of a superconducting transition on twinning. Only one series of experiments on indium single crystals gave a result which was positive to some degree. Five samples were deformed at 0.48 and 1.7 K alternately in the *N* and *S* states. The state was changed every minute. It was found that in four of five samples the first twin appeared in the *N* state. Analysis of the deformation curves showed that the frequency of the jumps in the *N* state is 20–30% higher than in the *S* state. The total amplitude of the jumps per unit time is correspondingly higher, though the average amplitude of the jumps was independent of the state.

Experimental investigations of low-temperature twinning in niobium using electric effects [96] did not detect any difference in the amplitude of the signals during the deformation of the samples in the normal and superconducting states. Experiments on zinc single crystals were found to be very informative [166]. Samples with a special orientation were studied. In this orientation basal glide was geometrically forbidden, and twinning in the system $(10\bar{1}2)$ $[10\bar{1}1]$ and pyramidal glide in the system $(11\bar{2}2)[11\bar{2}3]$ were the allowed forms of deformation. Metallographic observations of the deformed samples showed the presence of twins, and below 77.3 K the twins become smaller, the density of twins increases, and the twins fill the volume more uniformly. As a result, the macroscopic deformation curve becomes smooth, which is very convenient for studying the influence of a superconducting transition on twinning. The dislocation structure of deformed samples was studied to estimate the contribution of pyramidal glide to ε. In the process, fine lines of pyramidal glide were observed. Their contribution to the total deformation is negligibly small, i.e. when zinc single crystals are compressed along *[0001]* plastic deformation is due mainly to twinning. These are the samples that were chosen as the object of investigation. The deformation was performed by compression in the temperature range 0.5–1 K in liquid ^{3}He with constant rates 2×10^{-5}–2×10^{-4} s^{-1}. In the experiments the sample was cooled below T_c=0.825 K and deformed with repeated superconducting transitions. It was found that the flow stress changes appreciably by $\Delta\tau_{SN}$ at superconducting transitions (Figure 28). Right up to deformations ~4% the quantity $\Delta\tau_{SN}$ is independent of the deformation and equals ~80 kPa at 0.5 K; this is less than the value of $\Delta\tau_{SN}$ with basal gliding (250 kPa [118]) but much larger than with pyramidal glide ($\Delta\tau_{SN} \leq 10$ kPa [118]). The result obtained means that in this case the change of the flow stress at a NS transition is due to its effect on the nucleation and motion of twinning dislocations.

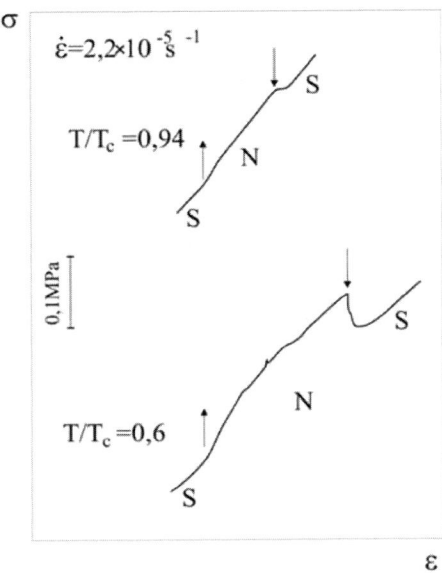

Figure 28. Influence of a superconducting transition on the flow stress with compression of zinc single crystals along [0001], when deformation occurs predominantly by twinning and is not accompanied by macroscopic disruptions of the load [120]

As one can see from the results presented above, a different effect of a superconducting transition on deformation twinning has been observed. Accordingly, several points of view have been advanced concerning the role of a superconducting transition on twinning.

Deformation twinning is sensitive to a superconducting transition indirectly, through the sensitivity of glide. This results in the appearance of different stress concentrators in the normal and superconducting states, resulting in a difference in the stresses at which twinning starts.

Twinning is sensitive to the electronic state of a sample to the extent of the change in volume at a superconducting transition. Then the sign of the effect should also depend on the sign of the volume change.

The sensitivity of twinning to a superconducting transition is determined by the rates of the process. The contribution of electronic drag is appreciable for low rates of twinning. For high rates, when breaking of Cooper pairs can occur, this is not the case. The nonuniqueness of the results obtained under twinning conditions distinguishes them from the results obtained under glide conditions.

VI. CORRELATION BETWEEN THE CHARACTERISTICS OF THE EFFECT AND THE SUPERCONDUCTING PROPERTIES

A number of experiments which provide convincing evidence of a correlation between the macrocopic characteristics of plastic deformation at a superconducting transition and the superconducting properties have been performed.

A. TEMPERATURE DEPENDENCE OF THE CHARACTERISTICS OF THE PLASTICITY CHANGE AT A SUPERCONDUCTING TRANSITION

1. Under the Conditions of Deformation at a Constant Strain Rate

The study of the temperature dependence of $\Delta\sigma_{SN}$ ($\Delta\tau_{SN}$) started immediately after the appearance of the first works, since it was important to understand the extent to which the plasticity change is associated with superconductivity. At first these were qualitative measurements,[2,30,40] where it was shown that at temperatures below $0.58T_c$ $\Delta\sigma_{SN}$ changes negligibly with temperature, at $T=0.47T_c$ and $0.53T_c$ $\Delta\sigma_{SN}$ is essentially temperature-independent, and $\Delta\sigma_{SN}$ decreases sharply as temperature increases up to $0.88T_c$. The first attempt to estimate analytically the form of the dependence $\Delta\sigma_{SN}(T)$ was made in Ref. 41. For this, a sample of polycrystalline indium was deformed at different temperatures. In so doing a correction not exceeding 10% was made for hardening. A linear relation was observed in the coordinates $\Delta\sigma_{SN}$ -$(T/T_c)^2$. In Ref. 43 it was pointed out that such an assertion is not definitive, and in Ref. 167 it is shown that the data of Ref. 41 can equally well correspond to a linear dependence of $\Delta\sigma_{SN}$ on T/T_c.

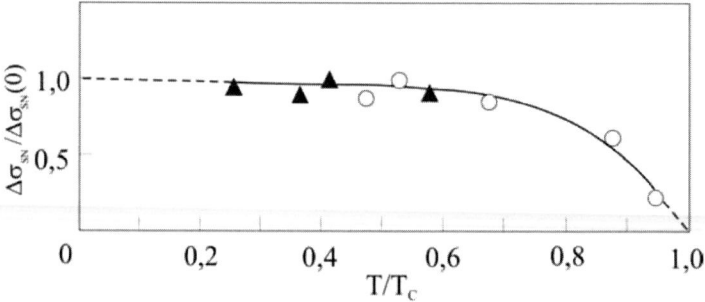

Figure 29. Temperature dependence of $\Delta\sigma_{SN}/\Delta\sigma_{SN}$ (0): (●) – свинец, (○)- индий [42]

An extensive experimental investigation with a large number of measurements on two superconductors with different T_c—polycrystals of lead (99.9995%) and indium (99.99%)—was performed in the temperature range 1.6–4.2 K to establish the character of the temperature dependence of $\Delta\sigma_{SN}$. This made it possible to cover a wide range of values of T/T_c (0.24–0.94) and to perform measurements on both metals in the interval (0.4–0.6). To represent the results in the form of a single curve (Figure 29) all measurements of $\Delta\sigma_{SN}$ were normalized to $\Delta\sigma_{SN}$ (0)—the jump in the stress at 0 K. As a result of such normalization, each point presented on the curve is an average of five or six values of $\Delta\sigma_{SN}(T)/\Delta\sigma_{SN}(0)$ obtained with deformations of 10, 20, 25, and 30%. In the interval $(0.24–0.6)T_c$ $\Delta\sigma_{SN}$ changes neglibily while for T/T_c=0.6–0.94 $\Delta\sigma_{SN}$ decreases sharply, going to zero as T_c is approached. The data obtained were found to be linear in the coordinates $\Delta\sigma_{SN}(T)/\Delta\sigma_{SN}(0)$ – (1-

$T/T_c)^{1/2}$. The energy gap of a superconductor near T_c exhibits the same temperature dependence— $\Delta \sim (1-T/T_c)^{1/2}$.

Subsequently, a series of works studying the dependences $\Delta\sigma_{SN}$ (T) and $\Delta\tau_{SN}$ (T) was performed for different metals and alloys with polycrystalline and single-crystalline structure. Aside from lead [2,30,40,42,45] and indium [41,42,85,86] aluminum [101,107], tantalum [99], lead-based alloys [54,122], aluminum-based alloys [107,154–156], zinc [101,118,120] and tin [111] were studied. A series of temperature dependences obtained by different authors for several materials are presented in Figures 30 and 31. Qualitatively, the character of the temperature dependence is the same in all cases. In the interval $(0.5T_c-T_c)$ $\Delta\sigma_{SN}$ ($\Delta\tau_{SN}$) sharply decreases as T_c is approached; below $0.5T_c$ it depends weakly on the temperature or is completely temperature-independent. The experiments performed on lead and aluminum alloys showed that the temperature dependence of $\Delta\sigma_{SN}$ ($\Delta\tau_{SN}$) is insensitive to the concentration of the alloying element. In many investigations attempts were made to link the results obtained to a concrete temperature dependence of some characteristic of superconductivity. Such characteristics are: the energy gap width Δ, which is proportional to $(1-T/T_c)^{1/2}$, and the critical field H_c, which is proportional to $1-(T/T_c)^2$; the density ρ_S of superconducting electrons in the BCS theory and in the Casimir–Gorter two-fluid model, proportional to $1-(T/T_c)^4$ and, the ratio of the ultrasonic absorption coefficients in the superconducting α_S and normal α_N states

$$\Gamma = \frac{\alpha_S}{\alpha_N} = \frac{B_S}{B_N} = \frac{2}{1+e^{\frac{\Delta(T)}{kT}}},$$

where B is the coefficient of dynamic electronic drag of dislocations. All theoretical temperature dependences are qualitatively similar. The existing experimental data taken together fall mainly between $\rho_N(T)/\rho_S(0)$ and $\Delta(T)/\Delta(0)$, which is evident in Figure 30. In many cases the experimental data are close to the temperature dependence of $1-\Gamma(T)$. As shown in Sec. IV D, the observed nonmonotonic temperature dependences of $\Delta\sigma_{SN}$ ($\Delta\tau_{SN}$) in certain alloys were due to magnetic flux trapping with repeated superconducting transitions produced with a magnetic field.

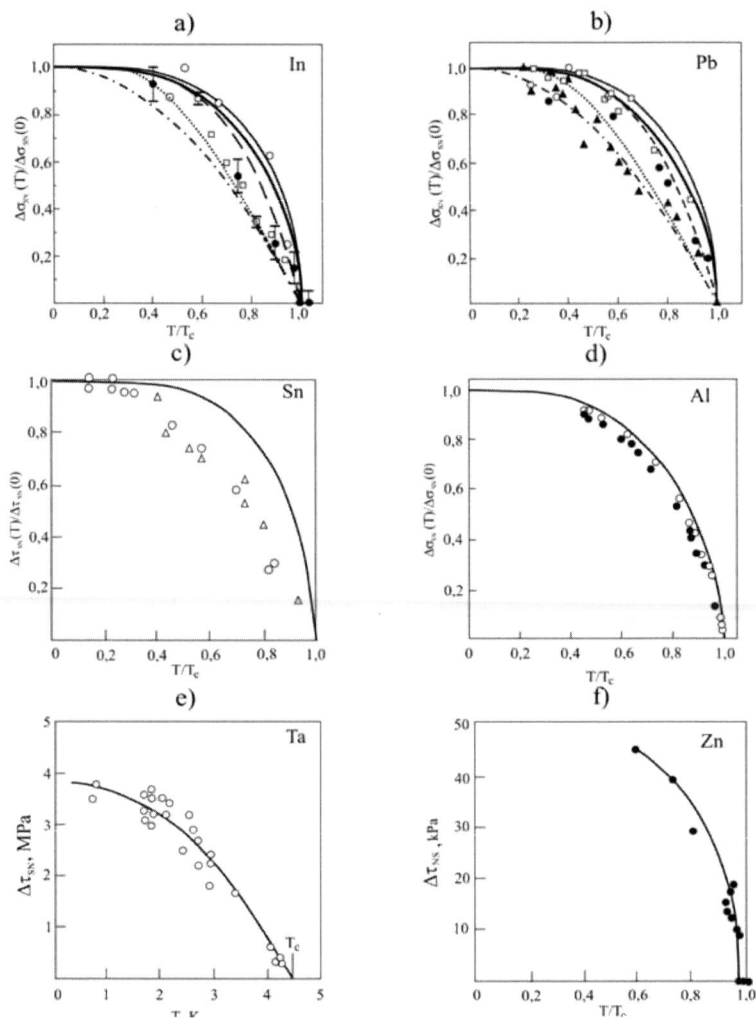

Figure 30. Temperature dependences of $\Delta\sigma_{SN}$ ($\Delta\tau_{SN}$) obtained on pure metals by different authors: indium (●)[41], (○) [42], (□) [95], $\Delta(T)/\Delta(0)$ (—); $(1-\Gamma)$(- - -); $1-(T/T_c)^4$ (---); $\Lambda(T)\Lambda(0)$(·····) (a); lead (○) [42], (●) [163], (□) [45], (▲) [45], $1-(T/T_c)^2$ (—·—) (b); tin [111] (deformation at constant rate) (○) [52]; (creep) (Δ),$\Delta(T)/\Delta(0)$(c); aluminum [101],—technical grade polycrystals (○) [101]; technical grade single crystals (●) [101]; $1-\Gamma$ (—)(d); tantalum, high-purity single crystals (RRR-700-12000). The data refer to different degrees of purity and different elongation from 1 to 15% (e) [99]; zinc, high-purity single crystals (RRR-1000] (f) [120]

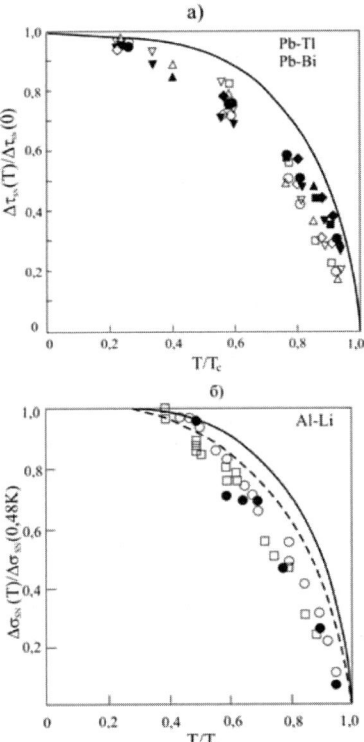

Figure 31. Temperature dependences of $\Delta\sigma_{SN}$ ($\Delta\tau_{SN}$) obtained for aluminum and lead alloys by different authors: pure lead (○, ●), lead—0.04 at.% Tl (△, ▲); lead—0.4 at.% Tl (▽,▼); lead—0.04 at.% Bi (□, ■); lead— 0.4 at.% Bi (◊,♦) [12]; (1-Γ)(—)(a); aluminum—3.8 at.% lithium (○), aluminum—7 at.% lithium (●), aluminum—10.4 at.% lithium (□), Δ(T)/Δ(0)(—); 1-Γ (---)(b) [156]

2. Under Creep Conditions

The qualitative study of the dependence $\Delta\varepsilon_{NS}$ (T) for lead [73] and indium [84] showed that $\Delta\varepsilon_{NS}$ decreases as T_c is approached and equals zero near T_c. In addition, a nonmonotonic temperature dependence $\Delta\varepsilon_{NS}$ (T) was observed for lead: below 2.5 K (0.35T/T_c) $\Delta\varepsilon_{NS}$ starts to decrease. A similar decrease of $\Delta\varepsilon_{NS}$ (T) was not observed in indium.

The temperature dependence $\Delta\varepsilon_{NS}$ (T) was measured in greater detail in Ref. 76 for polycrystals and single crystals of indium and lead with purity 99.999%. When the temperature dependence was studied, the experiments were performedat the nearly steady-state stage of creep in the stress range where $\Delta\varepsilon_{NS}$ is independent of the stress (see Sec.IV C). This made it possible to perform measurements on the same sample at different temepratures. Three or four measurementsof $\Delta\varepsilon_{NS}$ with loading were performed at the samevalue of the temperature and the average value was determined.The data from measurements performed on two samples of lead in the interval $(0.2-0.5)T_c$ and four samples of indium in the interval $(0.5-1.0)T_c$ were used to constructa single curve of $\Delta\varepsilon_{NS}$ (T)/ $\Delta\varepsilon_{NS}$ (0) versus T/T_c (Figure 32). The experimental temperature dependence $\Delta\varepsilon_{NS}$ (T) is monotonic, incontrast to Ref. 73, and differs quite strongly from the temperaturedependence of the energy gap Δ in a superconductorbut is close to the dependence $1-(T/T_c)^2$. A detailed study of the temperature dependence of $\Delta\varepsilon_{NS}$ in a wide interval of temperatures and deformations (right up to failure of the sample) was performed in Ref. 57. The temperature dependence of the ratio of the rates $\dot{\varepsilon}$ in the S and N states near a NS transition was determined at the same time. The experiments were performed on high-purity (99.999%) single crystals of lead in the interval (1.6−7.2) K. It was shown that the temperature dependences $\Delta\varepsilon_{NS}$ (T)/ $\Delta\varepsilon_{NS}$ (0) and $ln(\dot{\varepsilon}_S/\dot{\varepsilon}_N)_T / ln(\dot{\varepsilon}_S/\dot{\varepsilon}_N)_{T=0}$ are nonmonotonic and differ appreciably for different degrees of deformation, i.e. they are determined by the defect structure of the crystal. In Ref. 168 an attempt was made to determine the influence of tin impurities (1 and 3 at. %) on the temperature dependence of $\Delta\varepsilon_{NS}$ of a lead single crystal. In the alloys studied, strongly pronounced nonmonotonicity of $\Delta\varepsilon_{NS}/\Delta\varepsilon_{NS}$ (0) was observed below $T/T_c=0{,}7$. However, subsequently, it was shown that this is mainly due to the the trapping of magnetic flux. A monotonic temperature dependence of $\Delta\varepsilon_{NS}/\Delta\varepsilon_{NS}$ (0) was observed in lead with the paramagnetic impurity nickel [151].

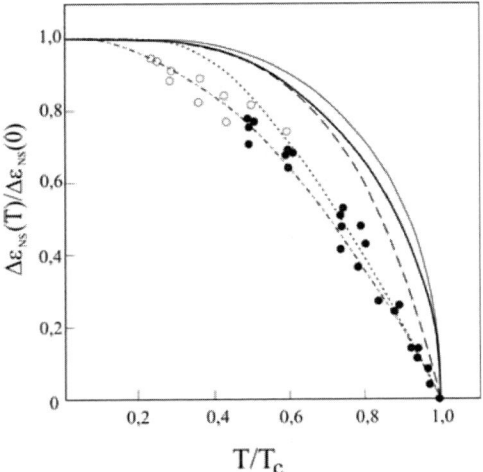

Figure 32. Temperature dependences of $\Delta\varepsilon_{NS}/\Delta\varepsilon_{NS}(0)$ for In(●) and Pb (○) [76], taken under the creep condition. Theoretical temperature dependences of superconducting parameters: (——) $\Delta(T)/\Delta(0)$; (——) $1-\Gamma$; ---- $1-(T/T_c)^4$; (········) $\Lambda(T)\Lambda(0)$; (–·–·–·) $1-(T/T_c)^2$

3. Under Stress-Relaxation Conditions

The dependence $\Delta\tau_{NS}^R(T)$ was initially studied in Ref. 39 on single crystals of high-purity lead (99.9999%), oriented for easy glide (the stretching axis is close to [110]). The comparison, made by the authors, of the temperature dependence obtained with the temperature dependence of the energy gap of a superconductor and the density of superconducting electrons showed that there is no strict agreement with any of these dependences, though the experimental data are closest to the temperature dependence of the gap (Figure 33). Next, the same authors [37] measured the temperature dependence of $\Delta\tau_{NS}^R$ for lead single crystals whose stretching axis was close to [100], and they obtained a dependence different from the one obtained for single crystals with easy glide [33], closer to the temperature dependence of the density of superconducting electrons. For alloys of lead with indium, bismuth, thallium, and cadmium (low concentrations) the dependence $\Delta\tau_{NS}^R(T)$ was found to be the same as for lead oriented for easy glide. Conversely, for high concentrations the dependence $\Delta \tau_{NS}^{R}$(T) is closer to that observed for pure lead with orientation close to *[100]*.

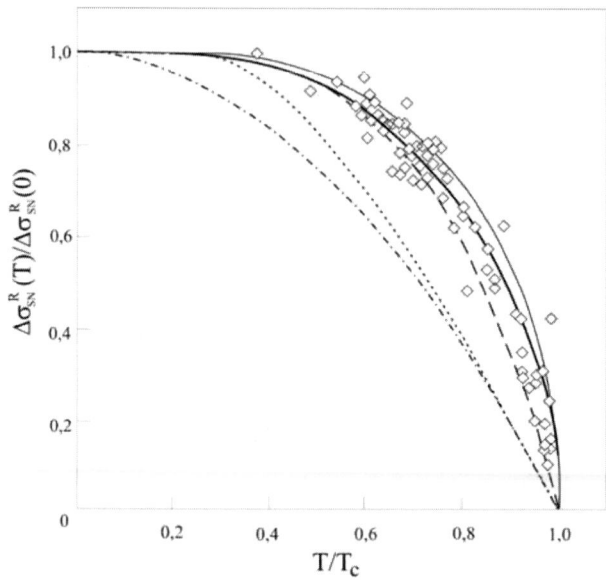

Figure 33. Temperature dependence of $\Delta \tau_{NS}^{R} / \Delta \tau_{NS}^{R}(0)$ for single crystals Pb-1 at. % Sn in the process of stress relaxation. Theoretical temperature dependences of superconducting parameters: (———) $\Delta(T)/\Delta(0)$; (———)1-Γ; ----1-$(T/T_c)^4$; (········)Λ(T)Λ(0);(— · — · —)· 1-$(T/T_c)^2$

Summarizing the existing experimental data, the following can be stated.

1. The temperature dependences of the characteristics of the change in plasticity at a superconducting transition are largely monotonic.

2. A characteristic feature of these dependences is the qualitative correspondence with the temperature dependences of the characteristics of superconductivity.

3. A quite large number of experimental data are close to the temperature dependence of 1-Γ(T), where Γ is the ratio of the ultrasonic abosrption coefficients in the *S* and *N* states or the ratio of the coefficients of dynamical drag of dislocations.

At the present time a more accurate correspondence between the experimental data and the theoretical curve cannot be obtained.

B. Magnetic Field Dependences of the Characteristics of the Effect

In the vast majority of the experiments described above superconductivity was suppressed by a magnetic field above the critical field of the superconductor. At the same time a magnetic field can itself influence the flow stress. To prevent possible magnetic effects the method of repeated loading (see Sec. III B 1) was used for one lead single crystal with stretching axis near [110] to determine the CRSS for different values of the magnetic field which was switched on [30]. The CRSS remained virtually unchanged in the interval 0–486 Oe (below H_c). The CRSS also did not change between the fields 936 and 1400 Oe (above H_c). However, the values of the CRSS above and below H_c differed from one another by 45% on the average. The CRSS changed between 468 and 936 Oe, where the value of H_c for lead lies. A more detailed experiment was performed on polycrystalline lead [30], where a magnetic field with different intensities was switched on and off in the course of continuous stretching of the sample. For fields below 530 Oe (H_c for lead at 4.2 K) at any stage of the deformation curve the field had no effect on the deforming stress, and fields above H_c caused σ to change by the amount $\Delta\sigma_{SN}$; the values of $\Delta\sigma_{SN}/\sigma$ for fields from 560 to 1870 Oe fell on the same curve of $\Delta\sigma_{SN}/\sigma$ versus the deformation δ, i.e. above H_c the magnetic field has no effect on $\Delta\sigma_{SN}$. In Ref. 1 superconductivity in polycrystalline lead at 4.2 K was suppressed by fields 1930 and 4790 Oe, the latter exceeding H_c by almost an order of magnitude, but no effect of a field on $\Delta\sigma_{SN}$ was observed. The same observation was made in Ref. 41 up to 4000 Oe and in Ref. 4 up to 6100 Oe. To determine whether or not a magnetic field in itself influences the flow stress, experiments were also performed at different temperatures. For this, during the deformation the sample temperature was raised gradually and a magnetic field with intensity above H_c was switched on and off. It is evident (Figure 34) that a change in the flow stress is observed only at temperatures below T_c. For $T>T_c$ switching a magnetic field on and off had no effect on the deformation curve. Thus, the effect of a magnetic field is observed when it changes the electronic state of the sample.

Since superconducting materials are divided into types and II, as determined by their behavior in a magnetic field in a definite interval of fields and the specific magnetic structure, it was of interest initially to make a comparative study of the characteristics of the change of the flow stress at *SN* and *NS* transitions in type-I and -II superconductors. The alloy In–3.9 at. % Pb (T_c=3.64 K) was found to be very convenient for such studies. For this alloy measurements of the magnetization curve and thermal conductivity [169] showed the existence of a different type of superconductivity in different temperature ranges—it is a type-II superconductor for $T>2.25$ K and a type-I superconductor for $T<2.25$ K. Two different types of superconductivity can be obtained in the same sample by changing the experimental temperature. The changes in σ at SN and NS transitions for temperature intervals corresponding to the different types of superconductivity are qualitatively the same and are similar to pure metals (Figure 35). The specific nature of the type of conductivity is manifested in the dependence of the parameters of the plasticity change at a superconducting transition on the magnetic field in intermediate (type-I superconductor) and mixed (type-II cuperconductor) states; this is studied below.

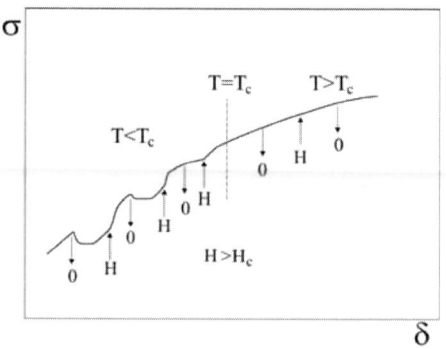

Figure 34. Section of the macroscopic tensile curve for lead with a magnetic field switched on and off repeatedly with intensity above the critical value at T<Tc, T>Tc, and T=Tc [45]

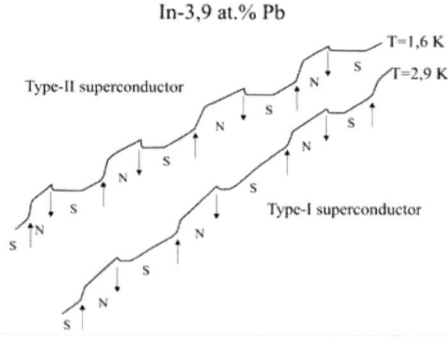

Figure 35. Portion of macroscopic tensile curves with repeated changes of the electron state of the polycrystalline In-3.9 at.% Pb sample. The curve was taken at different

temperatures: T=2.9 K (type-I superconductor) (a); T=1.6 K (type-II superconductor) (b) [169]

1. Intermediate State

It has been shown experimentally that the suppression of superconductivity in type-I superconductors occurs in a narrow range of fields. The presence of an interval of fields is due to the finite dimensions of the sample, where an intermediate state arises in fields somewhat below H_c. A collection of alternating normal and superconducting layers forms in this state. The relative fraction of the latter layers will increase as H_c is approached. The interval of fields corresponding to the intermediate state can be increased by changing the geometry of the samples and decreasing sample size. The experiment was performed on 99.9999% pure lead at 4.2 K.79 In such a sample (20×10×10 mm) under compression with rate 1.5×10^{-6} s^{-1} $\Delta\tau$ was observed starting at fields close to $0.8H_c$. For smaller samples finite values of $\Delta\tau$ are already observed at $0.5H_c$. As the intensity of the magnetic field increased to H_c $\Delta\tau$ was found to increase; this agrees with the field dependence of the magnetization of the sample. hysteresis of $\Delta\tau$ is observed as the field decreases below H_c; this also agrees with the hysteresis of the magnetization of a superconductor. The result obtained means that the field dependence of $\Delta\tau$ below H_c is due to the presence of an intermediate state of the sample.

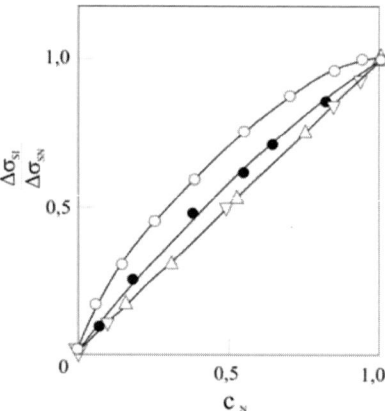

Figure 36. Variation of the flow stress $\Delta\sigma_{SI}/\Delta\sigma_{SN}$ of lead single crystals as a function of the concentration of the normal phase CN: $H\parallel$ (Δ,∇); $H\perp$ (●, ○); ratio of width to thickness of the sample: 1 (∇,●); 3 (Δ,○) [66]

Subsequently, detailed studies were performed in static and dynamic intermediate states. In Ref. 66 jumps in the flow stress at a transition from the superconducting state into the intermediate state ($\Delta\sigma_{SI}$) and the normal state ($\Delta\sigma_{SN}$) were measured as a function of the magnitude and direction of the magnetic field H relative to the stretching axis of the sample for polycrystalline 99.9999% pure lead at 4.2 K. The concentration C_N of the normal phase was determined at the same time according to the magnitude of the magnetic induction B and the resistivity ρ. It was shown that for H oriented parallel to the stretching axis $\Delta\sigma_{SN}$ is directly proportional to C_N (Figure 36), and for perpendicular orientation $\Delta\sigma_{SI} = C_N \Delta\sigma_{SN} + \Delta\sigma_b$, where $\Delta\sigma_b$ is the additional change in σ due to the interaction of the mobile dislocations with the interphase boundaries in the intermediate state.

The work presented in Ref. 66 was elaborated for lead polycrystals and single crystals at a transition from the superconducting into the intermediate state and the normal state as a function of the concentration of the normal phase, the sample size, the grain size, and the impurity concentration [80]. An additional confirmation of the role of interphase boundaries in dislocation drag in the intermediate state is the observed sensitivity of $\Delta\sigma_{SI}/\Delta\sigma_{SN}$ to the sample thickness, which determines the number of interphase boundaries. In Ref. 68

$\Delta\tau_{SI}$ in the intermediate state was also studied for single crystals of high-purity lead. The field dependences of $\Delta\tau_{SI}$ in a longitudinal field were qualitatively the same as those obtained in Ref. 66; a difference was observed in a transverse field. At 1.4 K in a transverse field the starting values of $\Delta\sigma_{SI}$ were negative. The quantity $\Delta\sigma_{SI}$ assumes positive values, which increase up to $\Delta\sigma_{SN}$ at H_c, only with a further increase of the transverse field. In Ref. 170 the contributions of the normal electrons and interphase boundaries of the static intermediate state of a single crystal and polycrystal of indium (99.9996%) in the temperature range 1.7–3.4 K in an external magnetic field with different orientation ($H^{||}$ and $H^{\perp}$) are distinguished. For a cyclic change of state in the course of stretching with rate 5 $\times 10^{-5}$ s^{-1} the sample was demagnetized to the initial state by short-time heating with electric field pulses. An increase of the flow stress by $\Delta\tau_{SI}$ in a single crystal and by $\Delta\sigma_{SI}$ in a polycrystal (grain size ~1 mm) is observed at a transition from the S state into the intermediate state with increasing field (H/H_c)<1).

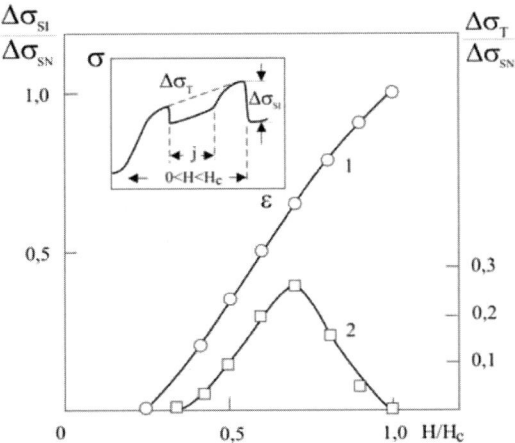

Figure 37. Variation of the flow stress of polycrystalline lead in a dynamical intermediate state as a result of the application of a magnetic field ($\Delta\sigma_{SI}$) and flow of electric current ($\Delta\sigma_{ST}$); $\Delta\sigma_{SI}/\Delta\sigma_{SN}$ (1) and $\Delta\sigma_{ST}/\Delta\sigma_{SN}$ (2) versus the applied magnetic field [171]

Analysis of the dependences of $\Delta\tau_{SI}/\Delta\tau_{SN}$ and $\Delta\sigma_{SI}/\Delta\sigma_{SN}$ on the magnetic induction B/B_c showed several features. There is no linear relation between the jumps $\Delta\sigma_{SN}$, $\Delta\tau_{SN}$ and the concentration of normal electrons $C_N=B/B_c$. The quantities $\Delta\tau_{SN}$ and $\Delta\sigma_{SN}$ in the field $H^\perp$ are always higher for the same values of B/B_c than the analogous quantities in the field $H^\parallel$. For the same magnetic field intensity, the relative quantity $\Delta\sigma_{SI}/\Delta\sigma_{SN}$ in a polycrystal is higher than in a single crystal. These features, together with the temperature dependences of $\Delta\sigma_{SI}$ and $\Delta\tau_{SI}$, likewise attest to the fact that in the intermediate state, aside from purely electronic drag of dislocations, additional drag occurs at the interface between the normal and superconducting phases.

In contrast to a static magnetic structure, the transmission of an electric current through a type-I superconductor in the intermediate state results in drift of the interphase boundaries. In Ref. 171 it was found that under the combined effect of a magnetic field H and an electric current with density j (H and j are perpendicular to one another and j is parallel to the stretching axis of the sample) in the intermediate state, aside from an increase of the flow stress by the amount $\Delta\sigma_{SI}$, switching on an electric current results in a decrease of σ by $\Delta\sigma_j$. This result (Figure 37), which is also presented in Refs. 89, 171, and 172 was another confirmation of the sensitivity of $\Delta\sigma_{SI}$ to the interaction of dislocations with a phase boundary in the intermediate state. In Ref. 128 careful measurements of $\Delta\tau_{SI}$ and the concentration of the normal phase in two type-I superconductors (pure lead and lead-0.5 at. % indium) were performed and the results obtained were analyzed taking account of not only the concentration but also the morphology of the normal phase.

2. Mixed State

Soon after it was discovered that the plasticity changes at a superconducting transition [1,2] the first experiments were performed to study the effect of a magnetic field on the flow stress of type-II superconductors [43,91,123,173]. In Ref. 91 experiments were performed on niobium single crystals oriented for single glide along the system $(\bar{1}01)[111]$. In the mixed state, starting at $0.4H_c$, the jump in τ increases (Figure 38,b) with the magnetic field directed along the stretching axis, reaching the maximum value $\Delta\tau_{SM} = \Delta\tau_{SN}$ at H_{c2}. The experimental values of $\Delta\tau_{SM}$ agreed well with the magnetic induction curve of the sample. As the field increases above H_{c2} $\Delta\tau_{SN}$ becomes independent of the field.

A similar experiment was performed on the alloy Pb–In [43,123,173]. The alloy Pb–13 at. % In was chosen for the investigations. In this alloy the mixed state occurs in a very wide range of magnetic fields (at 4.2 K H_{c1}=280 Oe and H_{c2}=2400 Oe). The polycrystalline samples were deformed at 4.2 K and were repeatedly transferred from the superconducting into a mixed state, and the magnetic field intensity was varied from H_{c1} to H_{c2} (Figure 39). The quantity $\Delta\sigma$ is zero in magnetic fields below H_{c1} and constant above H_{c2} (Figure 25a). In the mixed state $\Delta\sigma_{SM}$ agrees with the curve $B(H)$. Subsequently, similar experiments were performed on single crystals of the alloys Pb–4 at. % Tl, Pb–3.9 at % Bi, and Pb–0.85 at. % Cd [139] and polycrystals of Pb–4.6 at. % In, and Pb–13.2 at. % In [122], and on niobium [94,97]. Qualitatively the same result was obtained—$\Delta\sigma_{SM}$ ($\Delta\tau_{SM}$) increases in the mixed state as the magnetic field varies from H_{c1} to H_{c2}. Similar results were obtained in experiments in the mixed state under creep conditions (Pb–10 at. % In polycrystals) [123,135] and under stress relaxation conditions (single crystals of Pb–5 and 10 at. % In [37]; single crystals of Pb–5, 10, 15, and 20 at. % In [126] with one difference—the measurements of $\Delta\tau(\Delta\sigma)$ were performed only with the sample transferred from the mixed into the superconducting state; since flux trapping is possible in alloys this method can understate $\Delta\sigma_{SM}$ ($\Delta\tau_{SM}$).

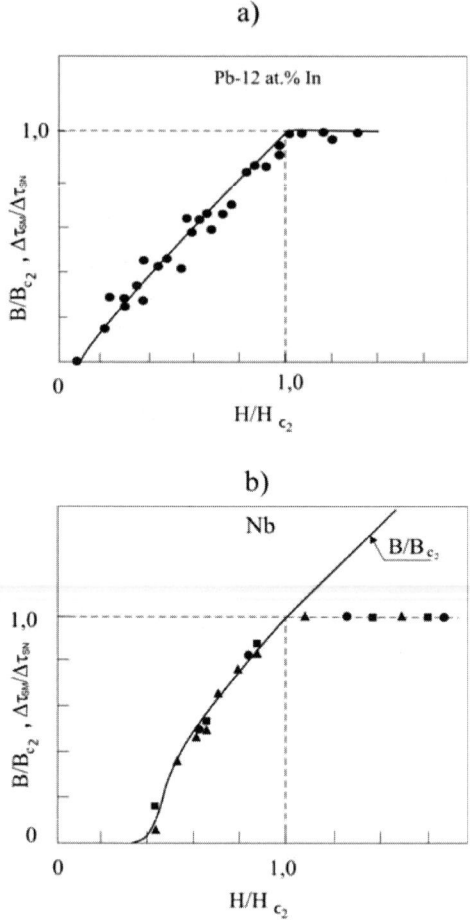

Figure 38. Variation of the normalized jump of the deforming stress $\Delta\tau_{SM}/\Delta\tau_{SN}$ (dots) in mixed and normal states as a function of the normalized magnetic field H/H_{c2}; for Pb–12 at.% In single crystals (a) [173]; for niobium single crystals (b) [91]. The line corresponds to the field dependence of B/B_{c2}

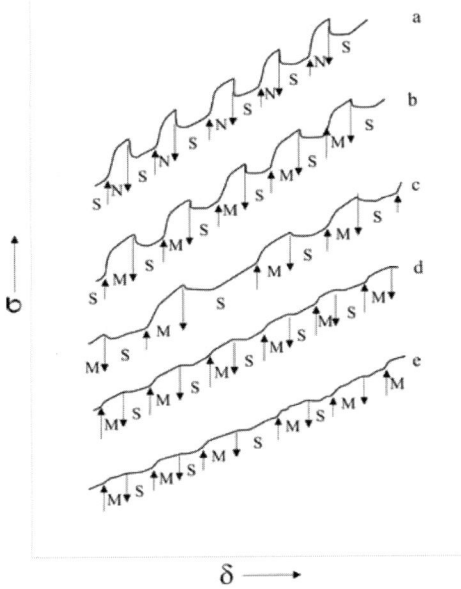

Figure 39. Portion of macroscopic tensile curves of polycrystalline Pb-12 at. % In with repeated changes of the electron state: H/Hc2=1.27 (a); 0.84 (b); 0.56 (c); 0.45 (d); 0.35 (e) [123, 173]

Comparative measurments in fields parallel and perpendicular to the stretching axis, performed on single crystals of the alloys Pb–2.2 and 12 at. % In [174,175], made it possible to observe the differences in the values of $\Delta\tau_{SM}{}^{\parallel}$ and $\Delta\tau_{SM}{}^{\perp}$. The concentration C_N of the normal phase was determined from the magnetic induction curves. It was established that aside from purely electronic drag, characterized by C_N, a term $\Delta\sigma_v$ which is a nonlinear function of CN appears in $\Delta\sigma_{SM}$. $\Delta\sigma_v$ is observed starting at $C_N \approx 0.4$, reaches its maximum value at $C_N \approx 0.7$, and decreases to 0 in the normal state. The most likely reason for the appearance of $\Delta\sigma_v$ is additional dislocation drag as result of the interaction of dislocations with magnetic field vortices. Apparently, the most careful measurements of $\Delta\sigma_{SM}$ ($\Delta\tau_{SM}$) and the concentration of the normal phase in the sample in the mixed state were performed in Ref. 130, so that this work merits a detailed exposition. The measurements were performed at 4.2 K on polycrystals and single crystals of lead-indium alloys with In concentrations up to 50 at %. The magnetic field was increased in steps from H_{c1} up to H_{c2} in a manner so as to obtain up to ten jumps of the flow stress $\sigma(\tau)$ and magnetic induction B, which were recorded simultaneously. This method differed substantially from the methods used in the works mentioned above. At the initial stages of the experiment the field was changed only in the direction of increasing values, which eliminated the errors due to magnetic flux trapping. The second advantage is that the jumps of $\sigma(\tau)$ and B were measured simultaneously, which made it possible to compare these quantities more accurately.

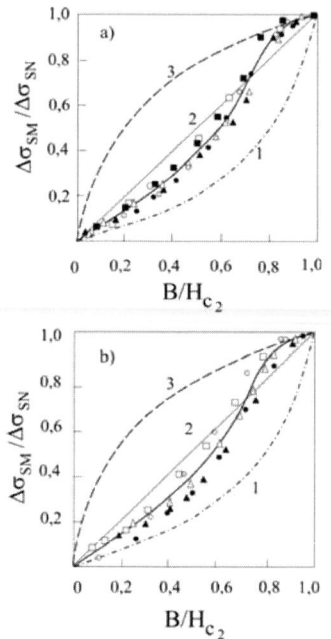

Figure 40. $\Delta\sigma_{SM} / \Delta\sigma_{SN}$ versus normal-phase concentration in type-II superconductors. Polycrystals of the following alloys: Pb–3 at.% In (increasing H(○); decreasing H (●)); Pb–5 at.% In (△); Pb–30 at.% In (▲); Pb– 50 at.% In (□); Pb–5 at.% In–5 at.% Sn (■) (a). Single crystals of the following alloys: Pb–2 at.% In (○); Pb–5 at.% In (●); Pb–10 at.% In (△); Pb–20 at.% In (▲); Pb–30 at.% In (□)(b) [130]

The dependences of $\Delta\sigma_{SM} / \Delta\sigma_{SN}$ for polycrystals and $\Delta\tau_{SM} / \Delta\tau_{SN}$ for single crystals on B/H_{c2} are strongly nonlinear and essentially identical for all alloys studied. A characteristic feature is inversion relative to the line of direct proportionality (see Figure 40), which is observed for C_N close to 0.7. This inversion was also observed in Ref. 81. The effect of H on $\sigma(\tau)$ vanished when the field reached H_{c2}. Analysis of the data obtained and comparison of these data with the results of investigations of the intermediate state showed that not only the concentration but also the morphology of the normal phase, which makes it possible to explain the observed nonlinearity and inversion, must be taken into account.

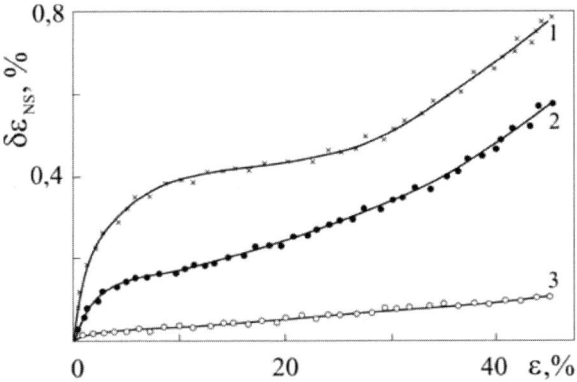

Figure 41. Deformation dependences of $\Delta\varepsilon_{NS}$ in polycrystals of a lead alloy with nonmagnetic impurities tin (1) and pure lead (2), and an alloy with the paramagnetic impurity nickel (3) [142]

C. EFFECT OF THE PARAMETERS OF THE ELECTRON ENERGY SPECTRUM

Another proof of the correlation between the observed plasticity changes and the superconducting properties are experiments studying the influence of a paramagnetic impurity on the additional creep $\Delta\varepsilon_{NS}$ at a superconducting transition [75,142,151]. Polycrystals of Pb–0.4 at. % Ni [142] were investigated under conditions of stretching at 4.2 K. The deformation dependence of $\Delta\varepsilon_{NS}$ is shown in Figure 41. This figure also shows for comparison the analogous curves for pure lead and the alloy Pb–0.4 at. % Sn. Evidently, nonmagnetic (curve *1)* and paramagnetic (curve *3)* impurities have the opposite effect on the dependence $\Delta\varepsilon_{NS}(\varepsilon)$ as compared with pure lead (curve *2)*. In the lead-tin alloy $\Delta\varepsilon_{NS}$ is 1.5–2 times greater than in pure lead, whereas in the alloy with nickel it is approximately an order of magnitude smaller than for pure lead. Special experiments showed that such changes are not associated with the difference in the plastic properties of the alloys. Consequently, the unusual influence of a paramagnetic impurity on $\Delta\varepsilon_{NS}$ was attributed to a strong decrease in the width of the energy gap. The same authors [151] expanded the interval of nickel concentrations (0.2, 0.4, 0.65 at. %) and measurement temperatures (1.6–7.2 K) and obtained similar results.

VII. Basic Theoretical Concepts of the Influence of a Superconducting Transition on the Plasticity

A. Electronic Drag of Dislocations in Normal and Superconducting States

The experimental observations of the change in the macroscopic characteristics of the plastic deformation of metals and alloys at a superconducting transition, which attest to an appreciable influence of conduction electrons on deformation processes, forced paying attention to, first and foremost, the few existing theoretical works which examined the interaction of moving dislocations with conduction electrons. These works were initiated by experiments studying the dislocation absorption of ultrasound and concerned the situation in a normal metal. Among existing publications, the works Refs. 18, 19, 177, and 178, performed for a rectilinear dislocation, are considered to be correct. The computed theoretical force of electronic drag of a dislocation in a normal metal is temperature-independent. Therefore the electronic drag coefficient B_e is temperature-independent; it was estimated to be 10^{-5}–10^{-6} dynes·s/cm². It is shown in Ref. 179 that in metals whose Fermi surface has a flattened section the electronic drag of dislocations with certain orientations contains a relaxation term proportional to the electron mean-free path, as a result of which B_e will depend on the temperature. In theoretical investigations of electronic drag of kinks on dislocations [180] it was found to be temperature independent. Since the expression for B_e contains the electron density (n_e), it was pointed out in Ref. 13

that at a superconducting transition B_e will change to the extent of the change in n_e.

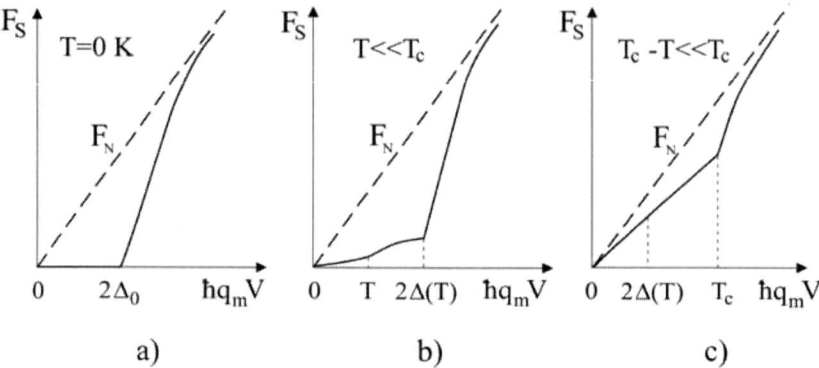

Figure 42. Schematic diagram of the electronic drag force FS on a dislocation in a superconductor versus the velocity V of the dislocation at various temperatures (solid curves) [182,183]; the dashed lines show the electronic drag force in a normal metal [18], qm—maximum momentum in the direction of motion of the dislocations

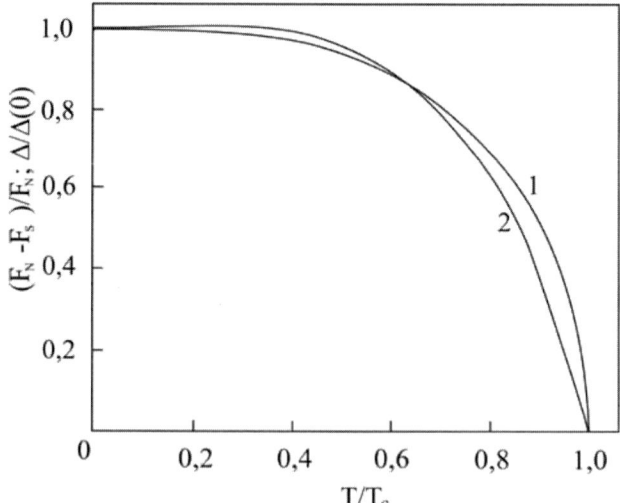

Figure 43. Temperature dependences of the relative gap width (1) and $(F_N - F_S)/F_N$ for $V \cong V_c$ (2), obtained numerically [185]

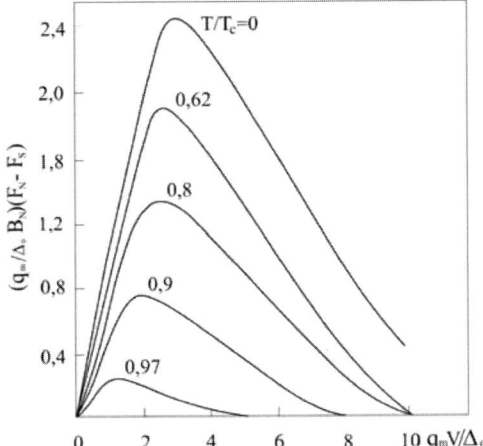

Figure 44. Difference of the friction forces FN-FS in the normal and superconducting states versus the dislocation velocity ($\omega m = q_m V$) at various temperatures, obtained numerically [185]

The first theoretical works which expressly studied the electronic drag of dislocations in the superconducting state were published in 1970–71 [181–183]. For dislocation velocities $V \sim 10^4$ cm/ s, which are easily achievable experimentally, the kinetic energy becomes comparable to the binding energy 2Δ of Cooper pairs. Consequently, a dislocation moving in a superconductor can give rise not only to transitions associated with scattering of "normal" excitations existing at a given temperature but also transitions associated with the creation of new excitations (breaking of Cooper pairs). As a result, the form of the velocity dependence of the electronic drag force F_S in the superconducting state changes substantially; this force varies with temperature, as shown for several limiting cases in Figure 42. As seen at 0 K, there exists a critical velocity V_c of a dislocation below which there is no electronic drag, and above which the drag increases charply as a result of the generation of quasiparticles. Taking account of umklapp processes [184] results in some smearing of the threshold in the velocity dependence of the friction force of a superconductor. The calculations performed in Refs. 182 and 183 made it possible to find the jump of the friction force $\delta F_{NS} = F_N - F_S$ at a superconducting transition. The temperature dependence of δF_{NS} in limiting cases is identical to the temperature dependence of the gap width $\Delta(T)$, but a unique linear dependence of δF_{NS} on $\Delta(T)$ does not exist. Figure 28 gives a qualitative idea of the character of the velocity and temperature dependences of F_S, and the quantitative characteristics, obtained using a computer, are presented in Refs. 181 and 185. The results obtained are valid for rectilinear and curvilinear dislocations with a radius of curvature which is much larger than atomic sizes. Consequently, the electronic drag on dislocation kinks, which control the plastic deformation of crystals with high Peierls barriers, requires a separate study. A detailed analysis of works on the electronic drag of dislocations is contained in the review Ref. 186 and the monograph Ref. 187. The influence of a superconducting transition on the motion of dislocations in a dynamical above-barrier regime can be estimated using the results of original works. In the general case $b\tau = F(\overline{V})$, where τ is the external stress and b is Burgers vector. In the normal state, at low temperatures $b\tau_N = F_N = B_N \overline{V}$, and in the superconducting state $b\tau_S = F_S(\overline{V},T)$. At a superconducting transition τ will be change to the extent of the change in F, $b(\tau_N - \tau_S) = b\Delta\tau_{SN} = F_N(\overline{V}) - F_S(\overline{V},T)$. Using the calculations for F_S, curves of ΔF_{NS} versus the external stress, temperature (Figure 43) and dislocation velocity (Figure 44) can be constructed [181–185]. The dependence of F_N–F_S on V and T is insensitive to the type (edge or screw) and form (closed or rectilinear) of dislocation [185]. Since most superconductors are metals which possess a complicated, multizone Fermi surface, the behavior of the electronic drag coefficient in superconductors with overlapping energy bands was studied in Ref. 188. It was shown that in two-band superconductors, just as in singleband superconductors, there exists a critical dislocation velocity, and the dependence of F_N–F_S on F_N is nonmonotonic. A purely dynamical regime of dislocation motion accompanying plastic deformation is quite rare, for example, at the head of the glide band under the action of large impulsive loads [27] and for high-rate deformation. In recent investigations dynamical effects were also observed under conditions of creep accompanying a superconducting transition [115,116] and under conditions of active deformation of concentrated lead-indium alloys [189]. An important feature of the dynamical drag of dislocations is its insensitivity to the degree of purity and doping. The experimental observations (see Sec. IV F) attest to a substantial dependence of the characteristics of the plasticity change at superconducting transitions on the concentration of the alloying element. These features and certain experimental data on plastic deformation at low temperatures (for example, Ref. 24) initiated the analysis of other mechanisms for the plasticity change at a superconducting transition.

B. FLUCTUATION MECHANISM

In the Refs. 190–192 it was proposed that the influence of electrons on the thermal surmounting of potential barriers, produced by impurity defects and dislocations of other systems, by dislocations be taken into account. This idea was developed in detail. The analysis is limited to crystals with negligibly small Peierls barriers (fcc and hcp crystals). It is supposed that the process of thermal activation is sensitive to the change in the electronic drag of dislocations at a superconducting transition. The idea that viscous drag influences the velocity of dislocations at low temperatures is advanced in Refs. 12, 193, and 194. This proposal is in agreement with the general principles of the theory of fluctuations, according to which the thermal fluctuations are associated with the dissipative properties of the medium. This was confirmed by an investigation of the thermal motion of a dislocation segment pinned by defects and by a calculation of the average rate of detachment from an individual defect. A modified Arrhenius equation, describing the kinetics of plastic deformation at low temperatures and low stresses, was obtained on the basis of this analysis. Using this equation and the dependence $B_S(T)$ (see Sec. VII A), expressions were obtained for $\Delta\tau_{SN}$, $\Delta\varepsilon_{NS}$ and $\Delta\tau_{NS}^R$. The fluctuation mechanism exhibits the following characteristic feature. The influence of a superconducting transition should be expected only for characteristic lengths of a dislocation segment $\sim 10^{-5}$–10^{-4} cm, which is characteristic of very pure metals. In Refs. 195–197 the fluctuation mechanism initially developed for a one-band superconductor was extended to the case of a two-band superconductor. The temperature dependence of $\Delta\tau_{SN}$ and $\Delta\varepsilon_{NS}$, just as in the case of a single-band superconductor, can be nonmonotonic. The concentration dependence of the parameters of the effect is also in qualitative agreement—it is a nonmonotonic function of the impurity concentration. Another possibility of the influence of viscosity on the fluctuation detachment of dislocations from local obstacles consists in taking account of the change in the effective temperature in the analysis of the quantum motion of dislocation segments on low temperatures [198]. A sharp decrease of the viscous drag coefficient B at a superconducting transition increases the effective temperature T^*, which increases the average rate of detachments from local defects and the rate of plastic deformation.

Subsequently, the theory of the fluctuation mechanism was developed for impurity crystals under conditions where quantum effects are effective, and nonmonotonic dependences were obtained for the parameters of the plasticity change at a *NS* transition on the impurity concentration and deformation [192]. The magnitude of the effect should reach a maximum at concentrations $\sim 10^{-3}$. The form of the dependence $\Delta\tau_{SN}(\varepsilon)$ depends strongly on the impurity concentration. The temperature dependence of $\Delta\tau_{SN}$ be nonmonotonic at definite low impurity concentrations.

C. INERTIAL MECHANISM

An inertial mechanism for dislocations to surmount a local barrier was proposed practically simultaneously with the mechanism described above [36,199–201]. At low temperatures the inertial mechanism becomes efficient and sensitive to a superconducting transition. The initial mechanism presumes that if a dislocation with effective mass M is moving with sufficiently high velocity, it can overcome an obstacle by means of inertial skipping. Such a situation can occur when dislocation segments go from a damped to an undamped state. A characteristic of damping is the expression

$$\mu = \frac{B\bar{l}V_s}{\pi G b^2} \quad (1)$$

where B is the dynamical drag coefficient, G is the shear modulus, b is Burgers vector, V_s is the sound velocity in the crystal, and $\bar{l}$ is the average length of a dislocation segment. If $\mu \geq 1$ (room and higher temperatures), dislocation segments are in a damped state and their inertial properties are not manifested. As temperature decreases, $\mu. < 1$ becomes possible because B and $\bar{l}$ decrease. In this case dislocation segments go from a damped into an undamped state in which their inertial properties can have a definite effect on the dynamics of dislocations in a crystal. If underdamped dislocation segments detach from a point defect (stopper) and, under the influence of an external load, strike a new stopper in the glide plane, then they skip through the position of static equilibrium (Figure 45) and for a short time $\sim 10^{-9}$–10^{-10} stay in a dynamical position, after which, having undergone several damped oscillations, they assume a new static equilibrium position. Therefore, as a result of its inertial properties, when a moving dislocation collides with a new stopper a dynamical increase by a factor of γ

($\gamma = tg(\alpha_{d,max})/tg(\alpha_{st})$) of the average angles under which a dislocation attacks the stoppers occurs. This can increase the average velocity of dislocations and therefore influence the macroscopic rate of deformation at low temperatures. For a number of years the inertial mechanism was a subject of theoretical investigations [8,10,36,47,139,194,199–207]. Recently, the motion of dislocations was simulated at an atomic level and observations were made of the inertial effects in a model experiment [206,207]. In these works it was proposed that the transition of dislocation segments from a damped into an undamped state should give rise to a corresponding transition from thermal activation to nonactivational viscodynamic motion of dislocations. However, in many cases the inertial properties of dislocations can be manifested with the thermal activation character of dislocation motion being preserved at the same time. The experimental data supporting this combination of thermal activation and inertial processes accompanying low-temperature deformation are summarized in Refs. 24 and 208.

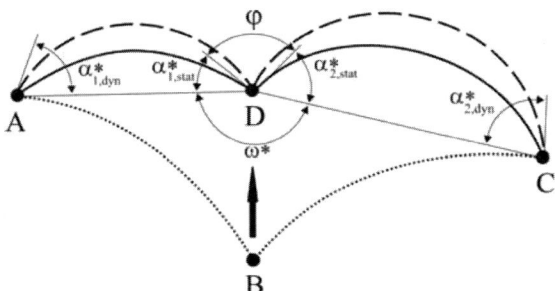

Figure 45. Diagram showing the manifestation of the inertial properties of an undamped dislocation line: dotted line—initial static position of a dislocation, resting on point defects (stoppers) A, B, and C; dashed line— dynamical position of a dislocation detached from the stopper B; solid line—new static position of a dislocation resting on stoppers A, D, and C [24].

D. THERMAL-INERTIAL MECHANISM

On the basis of the circumstances indicated in Sec. VII C attempts were made in a number of works to take account of the influence of the inertial properties of dislocations on the velocity of their thermal-activation motion. An analysis of the hypotheses advanced prior to 1987 is contained in Ref. 24. The problem of taking into account the influence of the inertial properties of dislocations on the velocity of their thermal activation motion turned out to be very difficult and remained unsolved for a long time. In 1979–1981 a method for solving this problem was proposed in Refs. 209 and 210. This method is based on a statistical description of dislocation motion in terms of a chaotic network of point obstacles. Analytic equations describing the average effective velocity $\overline{V}$ of thermal-activation motion of a dislocation and the corresponding rate $\dot{\varepsilon}$ of thermally activated plastic flow were formulated in these works. The inertial parameter γ appears explicitly in these equations. On this basis a thermal-inertial theory was developed and used to investigate the changes in the macroscopic characteristics of plastic deformation at a superconducting transition. We shall briefly examine the content of this theory. When dislocations move through a chaotic network of point obstacles the dislocation forms a series of successive configurations. It is assumed that all stoppers in the glide plane produce the same potential and force barriers whose height is characterized by a critical attack angle $\alpha_{\kappa p}$. The configurations found by a dislocation in the glide plane can be divided into two classes: mechanically stable, thermally activated configurations in which no attack angle exceeds $\alpha_{\kappa p}$ and mechanically unstable, nonactivational configurations, containing one or several angles greater than $\alpha_{\kappa p}$. The relative number of nonactivational configurations, among all the configurations which the dislocation successively forms as it moves through a chaotic network of point obstacles, is denoted as g. The quantity $g=0$ for $\sigma=0$ and $g=1$ for $\sigma \geq \sigma_{cr}$. Computer simulation showed that g is independent of temperature at low temperatures. For loads in the range $0.2\sigma_{cr} \leq \sigma < \sigma_{cr\_}$ the inequalities $0<g<1$ hold. Under these conditions the effective average velocity $\overline{V}$ of thermal activated motion of a dislocation through a chaotic network of local obstacles is described by the relation

$$\overline{V} \cong \frac{vS_0}{\overline{l}(\sigma)[1-g(\sigma,\sigma_{cr})]} \exp\left[-\frac{H_0}{kT}\ln(\frac{\sigma_{max}}{\sigma})\right] \qquad (2)$$

where v is the effective rate of thermal activation attempts; S_0 is the average area per stopper; $\overline{l}$ is the average length of a dislocation segment; H_0 is a parameter determining the effective enthalpy of thermal activation; $\sigma_{max} \geq \sigma_{cr}$, T is the temperature; and, k is Boltzmann's constant. The inertial properties of dislocations determined by the parameter γ do not affect the probability of thermal activational detachments of dislocations from stoppers, but they can increase the number g of nonactivational unpinning, i.e. $g(\sigma,\sigma_{cr})$ must be replaced by $g(\gamma^{3/2}\sigma,\sigma_{cr})$. It is this change that constitutes the method for taking account of the influence of inertial dislocations on the velocity of their thermal activation motion. The expression for $\Delta H(\sigma) = H_0 \ln(\sigma_{max}/\sigma)$ remains unchanged. Then, in general, Eq. (2) will have the following form taking account of the dynamical regime:

$$\overline{V} = \frac{\overline{\lambda}}{\overline{\lambda} v^{-1}_{\textit{дин,max}} + v^{-1} \exp[\Delta H(\sigma)/kT]} \qquad (3)$$

where in this case

$$\overline{\lambda} = \frac{1}{ac\overline{l}(\sigma)[1 - g(\gamma^{3/2}\sigma,\sigma_{кр})]} \qquad (4)$$

The relations (3) and (40 are the basic relations of the thermal-inertial theory proposed in Refs. 209 and 210. The relation $\dot{\varepsilon} = \rho b \overline{V}$ ($\dot{\varepsilon}$ is the deformation rate set by a machine and b is Burgers vector) is used to describe the macroscopic characteristics, assuming the density of mobile dislocations in the crystal to be constant. The physical results of the theory can be analyzed only by finding with the aid of a computer the numerical solutions of Eq. (2) for σ with certain chosen values of the parameters. The yield point τ_0 and its temperature dependence in the normal and superconducting state and, correspondingly, $\Delta\sigma_{SN}$ were determined in this manner. Figure 32 shows $\Delta\sigma_{SN}$ and the relative values $\Delta\sigma_{SN}/\sigma_{0N}$ as a function of the dimensionless concentration of stoppers $a/\sqrt{S_0}$ (a- is the lattice parameter) at fixed temperature. According to the theoretical results obtained, as the concentration of stoppers increases, the yield points σ_{0S} and σ_{0N}, increase, as a result of which $\Delta\sigma_{SN}$ increases. However, as $a/\sqrt{S_0}$ increases further, the average travel distance of dislocations $S_0/\overline{l}$

decreases to such an extent that there is not enough time for a dislocation to accelerate to the velocity $V_{d,max}$. As a result, the influence of a *NS* transition should weaken. A maximum on the concentration dependence of $\Delta\sigma_{SN}$ was first predicted theoretically in Ref. 202 and is confirmed by numerical calculations (Figure 46). Figure 33 shows the temperature dependence of $\Delta\sigma_{SN}$ following from the thermal-inertial theory. The theoretical temperature dependence $\Delta\sigma_{SN}(T)$, just as the experimental dependence (see Sec. VI A), is characterized by a monotonic increase of $\Delta\sigma_{SN}$ and by the fact that it goes to a constant value in the limit $T \rightarrow 0$.

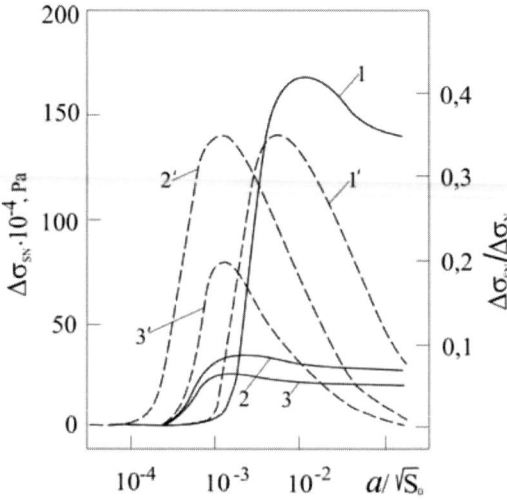

Figure 46. $\Delta\sigma_{SN}$ (solid lines) and the relative values $\Delta\sigma_{SN}/\Delta\sigma_{N}$ (dashed lines), following from the thermal-inertial theory for different values of the computational parameters (H0=0.1 eV; T=0.1Tc): $\alpha cr = \pi/6$, BeN=5×10−5 g·cm−1· s−1 (1) and (1'); $\alpha cr = \pi/6$, BeN =10−5 g·cm−1· s−1 (2) and (2'); $\alpha cr = \pi/46$, BeN=10−5 g·cm−1· s−1 (3) and (3') [24]

The experimental investigations of the low-temperature plasticity performed in the last few years [211–213] under conditions of active deformation and creep have confirmed the effectiveness of the thermal-inertial mechanism.

The theoretical investigations of the mechanism including thermal activation and inertial effects continued. A onedimensional model employing the equation of Brownian motion in application to the motion of dislocations is proposed in Ref.

214. The motion of dislocations in the twodimensional case through a random network of obstacles, which includes inertial detachment of underdamped dislocation segments, is examined in Ref. 215. In the form obtained these hypotheses are difficult for drawing comparisons between theory and experiment.

E. QUANTUM-INERTIAL MECHANISM

Anomalies of the plasticity occur in pure metals at low temperatures (below 10 K). These anomalies can be explained by the quantum properties of dislocations [216,217]. Consequently, strictly speaking, at especially low temperatures, aside from inertial effects and thermal activation, the quantum properties of dislocations must be taken into account. The result of the combined effect of quantum, inertial, and thermal fluctuation effects is a transformation of the classical Arrhenius equation [146]

$$\dot{\varepsilon} = \dot{\varepsilon}_0 \exp\left[-\frac{\Delta H(\tau^*)}{kT}\right] \tag{5}$$

($\dot{\varepsilon}$ is the plastic deformation rate; τ is the external stress; τ_i is the internal stress) into an equation of the following form:

$$\dot{\varepsilon} = \dot{\varepsilon}_0 I(\tau^*, B) \exp\left[-\frac{\Delta H(\tau^*)}{kT^*(T)}\right] \tag{6}$$

The temperature T in the denominator in the exponential in Eq. _5_ is replaced in Eq. _6_ by the function $T^*(T)$—the effective temperature of a dislocation taking account of its thermal and quantum motions [217]. The pre-exponential factor $\dot{\varepsilon}_0$ in Eq. (5) in front of the exponential is replaced by the function $\dot{\varepsilon}_0 I(\tau^*, B)$, which depends on the effective stress τ^* and the dynamic drag constant B. A simple approximate formulas has been obtained for the effective temperature $T^*(T)$:

$$T*(T) = \begin{cases} T, & T > \theta \\ \dfrac{\theta}{2}\left(1 + \dfrac{T^2}{\theta^2}\right), & T < \theta \end{cases} \qquad (7)$$

Here θ is the characteristic temperature of a dislocation, which is the main characteristic of its quantum properties. The function I reflects the effectiveness of the inertial effects, which was examined in Sec. VII D. In the theory of this mechanism expressions have not yet been obtained for the dependence of $\Delta\tau_{SN}$ on the deformation, temperature, and impurity concentration.

F. OTHER THEORETICAL HYPOTHESES

Several other hypotheses have be advanced. They should be mentioned to complete the picture. The experimental observation of an appreciable increase of creep at a *NS* transition at a stage where, within the limits of experimental accuracy, the deformation rate is close to zero [55], has given rise to the hypothesis of a possible quasistatic mechanism in which the characteristic of a potential barrier hindering dislocation motion changes. A quasistatic mechanism was proposed in Refs. 218 and 19. The basis of this mechanism is taking account of the nonuniformity of the electronic structure of a superconductor, which is related with the nonuniformity of the dislocation structure arising as a result of the heterogeneity of plastic deformation, as a result of which the internal stresses give rise to local changes of the energy gap Δ during deformation. The direct effect of dislocations on the phonon spectrum also results in local changes of the gap width. The nonuniform distribution of dislocations and point defects changes the electron mean-free path length, which locally changes the energy of the *NS* boundary. All this gives rise to spatial nonuniformity of the superconducting state and increases the free energy density in the *S* state by the amount $\delta F \sim N(0)\xi_0^2(grad\Delta)^2$, where $N(0)$ is the density of states at the Fermi surface and ξ_0 is the coherence length. The presence of a gradient of the free energy results in an additional thermodynamic force which strives to make the dislocation structure more uniform. The effective mechanical stresses arising as a result of this force are estimated to be 10^5 dynes/cm^2.

It is shown in Ref. 220 that the decrease of the activation energy as a result of a decrease of the potential barriers can be attributed to an increase of the free energy of the super- fluid component of the conduction electrons, screening the perturbation of the positive charge of the ions in the region of a deformation of the crystal lattice near a dislocation. This increase even exceeds the decrease of the potential barrier $\delta U = 0{,}4 \times 10^{-3}$ eV, which for lead is required to explain the observed value of $\Delta \varepsilon_{NS}$.

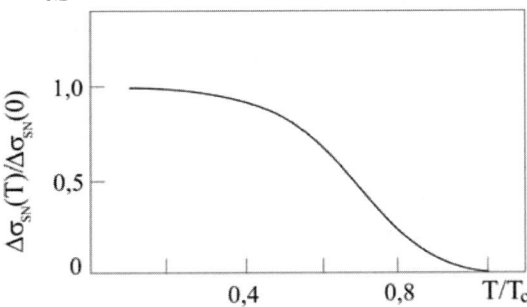

Figure 47. Temperature dependence of $\Delta \sigma_{SN}$ obtained in the thermal-inertial theory [24]

The possible role of the real structure of local barriers in estimating the magnitude of the effect of a superconducting transition is indicated in Ref. 221. Instead of the ordinarily studied motion of dislocations through local stress fields, corresponding to centers of dilatation in an isotropic medium, in crystals with strong anisotropy the complicated stress fields resulting in oscillations of the "dislocationdefect" interaction force should be taken into account.

A hypothesis explaining the change in plasticity at a superconducting transition without using the interaction of dislocations with conduction electrons has been advanced in Ref. 83. This hypothesis is based on the fact that because of a decrease of its electronic component the thermal conductivity of a metal in the superconducting state becomes less than in the normal state. The authors believe that this should decrease the rate of heat removal from the region of plastic deformation, which will result in local heating. In turn, the heating should increase the rate of thermally activated plastic deformation. Thus additional deformation due to a decrease of the flow stress should occur at a superconducting transition. According to Ref. 83, such a decrease of $\Delta \tau_{SN}$ caused by heating is expressed by

$$\Delta \tau_{SN} = -\beta A \ln \frac{K_N}{K_S} \qquad (8)$$

where $A = d\tau/dT$; and K_N and K_S are the thermal conductivities in the normal and superconducting states. In pure superconductors $K_N/K_S > 1$, resulting for $d\tau/dT < 0$ in positive values of $\Delta\tau_{SN}$; this is observed experimentally. However, situations where $d\tau/dT > 0$ are possible (low-temperature anomaly of the yield point and flow stress). In this case, according to Eq. (8), $\Delta\tau_{SN}$ should be negative. A similar result should be obtained for $K_N/K_S < 1$, which is characteristic for certain superconducting alloys. Negative values of $\Delta\tau_{SN}$ have not been observed experimentally in any material studied.

Special experiments on lead-bismuth alloys, in which $K_N/K_S < 1$, have shown [82] that $\Delta\tau_{SN} > 0$ and depends on temperature just as in superconductors with $K_N/K_S > 1$. True, the authors of the thermal hypothesis [83] believe that it is valid only for pure metals. In Ref. 82 experiments were also performed on pure lead in the intermediate state produced by a magnetic field transverse to the stretching axis of the sample. In this case $K_N/K_S < 1$, and the measured quantity $\Delta\varepsilon_{NS} > 0$. In addition, analysis showed that the thermal relaxation time is many orders of magnitude shorter than the time required by dislocations to move between barriers.

We mention two other works where hypotheses differing from those examined above are briefly advanced. In Ref. 222, in an analysis of the particular case of the influence of a NS transition on plasticity in pure lead, a model of strain hardening is proposed, where the deformation is viewed as a consequence of intermittent glides. A change of the dynamic drag coefficient B at a superconducting transition changes the number of intermittent glides. The quasistatic mechanism, based on taking account of the gradient of the energy gap of a superconductor near the core of an edge dislocation, is proposed in Refs. 223 and 224.

In [225, 226] it was consider a new mechanism for thermal-fluctuation motion of dislocations which makes it possible to allow for the quantum and dissipative effects in unified manner. The main conclusion derived from this model are as follows. In studies of the fluctuation-induced overcoming of a pinning center by a dislocation it is essential to allow dynamics of an obstacle due to its finite mass. Since the probability of overcoming of a light obstacle is greater than that of a heavy one, the plastic properties of a material are governed by the system of heavy obstacles. Experimentally observed temperature-independent anomalies are manifested only at low temperature because of the dominant role of

Plasticity of Metals and Alloys 83

heavy centers point out above. The characteristic temperature corresponding to the onset of quantum fluctuation in real crystals is governed by the frequency of a quasilocal mode of the obstacle and temperature-independent anomalies occur at the temperature on the order of tens of Kelvin. At sufficiently At sufficiently low temperatures the probability of fluctuation-induced overcoming of an obstacle is governed by dissipative effects also and depends exponentially on the viscosity determined by quasiparticle exitations.

Table 1. Comparison of existing theories with the experimental data on the influence of a superconducting transition on plasticity

Factors	Experimental data	Theories		
	Flow stress	Dynamical	Thermal-inertial	Fluctuation
Stress				
Temperature				
Strain rate				
Alloying				

G. COMPARISON OF THE THEORETICAL AND EXPERIMENTAL RESULTS

The theories can be compared with experiments by comparing the dependences of the parameters of the plasticity change at a superconducting transition on the stress (deformation), temperature, concentration of the alloying element, and strain rate, which are obtained experimentally and follow from the theoretical hypotheses. Such dependences are summarized in Table 1. This makes it possible to compare theory with experiment qualitatively and give preference to the fluctuation-inertial mechanism in which thermal and quantum fluctuations can be effective. The temperature dependences, which follow from the various mechanisms which have been worked out in detail, of the parameters of the effect are largely similar and close to the temperature dependence of the energy gap of a superconductor, just as the experimental data as a whole. This makes it difficult to give preference to any one mechanism on the basis of an analysis of the temperature dependences. It is difficult to use of the experimental rate dependences because there are so few of them. Preference can be given to the fluctuation-inertial mechanism on the basis of the qualitative agreement between the experimental and theoretical dependences of the parameters of the effect on the concentration of the alloying element and on the degree of deformation.

The fluctuation-inertial mechanism is also supported by the satisfactory explanation which it gives for the characteristics of low-temperature plasticity of metals [146,211–213,227], in which an influence of a superconducting transition on plasticity is observed.

H. Theoretical Hypotheses Concerning BCC Crystals

The analysis thus far has concerned fcc and hcp crystals, where plastic deformation is determined by the interaction of dislocations with point defects, and where the Peierls barriers are comparatively low. This is explained, first and foremost, by the fact that the overwhelming majority of the experiments studying the influence of a *NS* transition on plasticity have been performed precisely on these crystals. Superconductors with bcc structure (for example, niobium, tantalum, niobium-molybdenum alloys, molybdenum) have been studied experimentally much less, and correspondingly there are fewer theoretical hypotheses.

Initially, a theoretical estimate of the change in the kinetics of deformation at a superconducting transition in a crystal with Peierls mechanism of plasticity was proposed on the basis of an experimental study of the effect of a superconducting transition on the deforming stress of molybdenum single crystals [102,103]. The theory of thermal-fluctuation nucleation of a double inflection (kink) in the Peierls relief was used [216]. For the analysis it is important that the probability of nucleation of an inflection is proportional to the diffusion coefficient D of a single inflection on a dislocation, which depends on the electronic state of the superconductor; in addition, $D_S > D_N$. Comparing the experimental values of $\Delta\tau_{SN}$

for single crystals of highly pure molybdenum made it possible to determine the ratio of the diffusion coefficients of inflections in the N and S states: $D_S/D_N \approx 2$. Subsequently [112], theoretical expressions were obtained for all characteristics of the plasticity of a metal with a Peierls mechanism of glide at temperatures where quantum tunneling of dislocations becomes effective. On this basis an expression was proposed for $\Delta\tau_{SN}$. The detailed theoretical analysis performed in Ref. 112 of the experimental results obtained for tin single crystals in the normal and superconducting states qualitatively confirmed the prediction of the theory of quantum motion of damped dislocations in a Peierls relief.

The experimental data on the influence of a NS transition on the yield point and the rate dependence of the flow stress of highly pure single crystals of tantalum [33] were compared in Ref. 100 with a recent theory [228] in which a general calculation was performed of the rate of formation of pair inflections at any temperature. The calculation is based on the string model, taking into account dissipation with a generalization of the one-dimensional theory of quantum tunneling. According to this analysis the viscosity change occurring at a SN transition influences the rate of dislocation tunneling through a Peierls potential. In the opinion of the authors of Ref. 100 there is satisfactory quantitative agreement between the experimental data on $\Delta\tau_{SN}$ and the theory. True, it is noted in Ref. 100 that below 4 K (i.e. precisely below T_c) substantial discrepancies are observed between theory and experiment on measurement of the temperature dependence of the yield point. Such discrepancies could be due to measurement errors and imperfections of the theory.

VIII. DIRECT EXPERIMENTS

Many experimental studies have been published since the first works appeared. Most of these have been generalized in Secs. IV–VI. However, some of them, which are fundamental for understanding the mechanisms and dislocation processes occurring at a superconducting transition and in a superconducting state merit a separate analysis.

A. INFLUENCE OF THE SUPERCONDUCTING STATE ON THE MOBILITY OF SISLOCATIONS IN NB

The first, and so far only, successful investigation of dislocation mobility in the normal and superconducting states is Ref. 93. This is due to methodological difficulties in reliably revealing dislocations in a superconductor. To investigate mobility a dislocation was introduced into a single-crystal sample by pricking the surface with an indentor [140]. The dislocations were in the glide systems [110] ⟨111⟩. The sequence of experiments was as follows. At 300 K imprints were made uniformly along the entire, 25 mm long, lateral surface of a sample. Next, the sample was placed in a loading apparatus, cooled to 4.2 K, and loaded by means of three-point bending for $10-10^4$ sec. The stress created in this process varied literally from zero at the edges of the supports to 200 MPa at the center of the sample. If the experiment was performed in the normal state, the sample was located in a longitudinal field 7 kOe. After the load was applied the sample was heated to room temperature and the dislocations in the rosette pricks on the compression side were revealed by chemical etching.

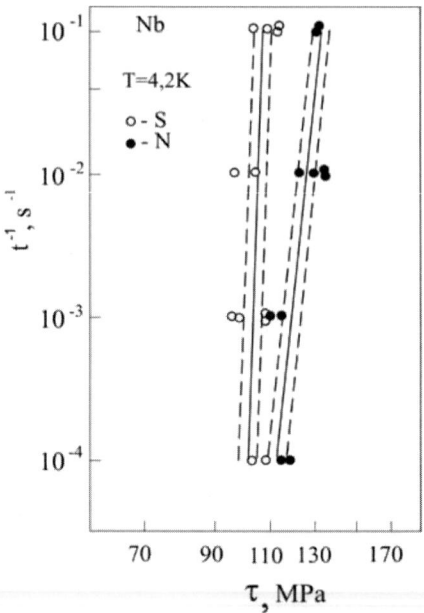

Figure 48. Data on dislocation mobility in N and S states in the coordinates lg t -1 -lgτ in niobium single crystals. The dashed lines show the confidence interval determined with reliability 0.9 for a series of experimental points in each state. T=4.2 K [93]

The stress τ of the onset of dislocation motion, corresponding to a prescribed loading time t, was determined. Only the cases where dislocation motion was observed in 70% of the rosettes were taken into account. The dependence $t^{-1}(\tau)$ obtained characterizes the dislocation mobility; t is the average passage time through the "difficult" position. The results are presented in the coordinates log t^{-1}–log τ in Figure 48. The measurements show an appreciably different mobility of dislocations. For $t^{-1} = 10^{-1}$ s^{-1} $\Delta\tau_{NS} = \tau_N - \tau_S = 25 \pm 5$ MPa. The slope m= $\Delta lg t^{-1} / \Delta lg \tau$, can be estimated from the curves $t^{-1}(\tau)$ obtained; the slope also differs: m_N=45±26 and m_S =130±100. The values obtained for m attest to thermal activation and not viscous motion of dislocations. On the other hand an estimate of dislocation velocities, assuming $V=L/t$ and L=const, gives V_S/V_N=7×10^4. This value is much greater than the theoretical estimates [186], according to which for niobium $V_S/V_N \approx 20$. The large value of V_S/V_N is attributed in Ref. 93 to inertial or thermal-inertial effects.

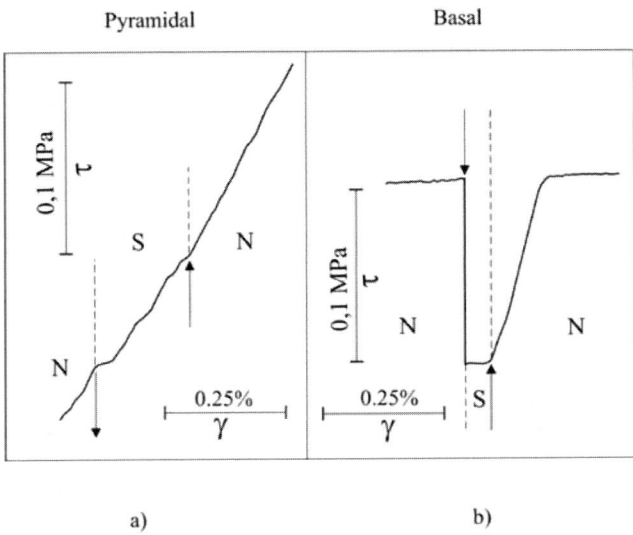

Figure. 49. Sections of hardening curves with compression of zinc single crystals with a change in the electronic state of a sample with pyramidal (a) and basal (b) glide. T=0.45 K; $\dot{\varepsilon}$=1.6×10^{-5} s^{-1} [118]

B. Study of the Influence of the Electronic State on Pyramidal and Basal Glide

The influence of a superconducting transition on the flow stress of zinc single crystals was studied in Ref. 118. The experiments were performed under conditions of compression with deformation rate 10^{-5}–10^{-6} s^{-1}. The 10×4×4 mm samples had two orientations of the compression axis. For one orientation only basal glide along the system $(0001)<11\bar{2}0>$ occurred. For the other orientation glide along the planes of a pyramid $\{11\bar{2}2\}<11\bar{2}3>$ was effective. The initial density of basal dislocations was 1–3×10^5 cm^{-2} and the density of pyramidal dislocations was 2×10^3 cm^{-2}. The experiments were performed at 0.45 K. Figure 49 shows sections of the hardening curve with NS and SN transitions in the case of pyramidal (a) and basal (b) glide. It is evident that under the conditions of basal glide an appreciable jump $\Delta\tau_{NS}$ is observed in the flow stress. The form of this jump is similar to the analogous jumps in other plastic and more isotropic, from the standpoint of glide, metals and alloys. In contrast to this, the quantity

$\Delta\tau_{NS}$ is vanishingly small under conditions of pyramidal glide. This difference is probably due to the substantially different mobility of basal and pyramidal dislocations [230,231]. For basal dislocations a high velocity and viscous motion are characteristic, whereas for pyramidal dislocations thermal activated and viscous motions are observed in the mobility. It can be concluded on the basis of the data obtained that the intensity of the manifestation of a *NS* transition in plasticity is due primarily to the mobility of dislocations and not the change of their density. Experiments under conditions of basal glide on crystals with different dislocation density of a "forest" (from 2×10^3 cm^{-2} up to 10^6 cm^{-2}) attest to this. As density increases, $\Delta\tau_{NS}$ decreases by an order of magnitude. This is due to a transition from viscous to thermally activated dislocation motion.

C. Displacement of Dislocation in the Normal and Superconducting States

The experimental procedure consisted of the following [166]. A zinc single crystal was cooled to $T<T_c$ and scratches were made with an indentor on the surface of the sample at constant temperature. During scratching the sample was transferred from a superconducting into the normal state and vice versa by switching a magnetic field on and off several times. After being heated to room temperature the sample was etched in order to reveal the dislocations introduced by scratching, and their displacement in the *N* and *S* states were measured. The displacement of basal, pyramidal, and twinning dislocations were studied. The displacement of basal dislocations was found to be sensitive to the electronic state of the sample. The displacement of basal dislocations in the load range 0.06–0.27 N are shown in Figure 50, where each point corresponds to an average of the value of *l* over 100–240 measurements. As the load on the indentor increases, the average displacement increases, and $l_S>l_N$ by 15–40% always holds. Under the smallest load 0.104 N there are virtually no runs of the pyramidal dislocations. Then, as the load increased to 0.8 N, the average displacement increased monotonically, and $l_S>l_N$ by 10–30%. Just as for motion of perfect dislocations, the displacement of the leading twinning dislocations increases with the load on the indentor, and the average length of twins for different velocities of the indentor in the superconducting state is 15–20% larger than in the normal state in the entire load range. The indefiniteness of the load and its application time makes it difficult to convert correctly from the ratio of the average displacement $\bar{l}_S/\bar{l}_N$

to the velocity ratio V_S/V_N, which is necessary in order to compare the experimental data with the theories. If it is assumed that for a constant rate of motion of the indentor the load application time was the same in both states, then it can be assumed that $V_S > V_N$ by a factor of 1.1–1.6.

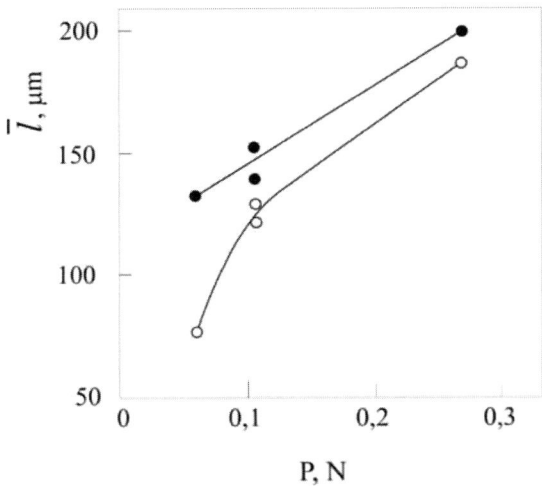

Figure 50. Travel distances of basal dislocations in zinc single crystals from a scratch made in S (●) and N (○) states. P—load on indentor, velocity of indentor 1 mm/min, T=0.5 K [166]

D. Defect Structure of a Crystal after Deformation in the Normal or Superconducting States

The defect structure arising in a crystal after deformation in a different electronic state has been studied in several works by different methods. In Ref. 231 the resistivity R of polycrystalline lead (99.9994%), deformed at the transient stage of creep at 4.2 K in the normal and superconducting states, was measured. The sequence of operations was as follows: a) R was measured in the initial normal state; b) a load was applied to the sample in steps in the normal state and R was measured during the creep process; c) the sample was transferred into a superconducting state, and in the process additional creep occurred; d) after creep for one minute in the S state the sample was transferred into the N state to measure R. Identical deformations in the normal and superconducting states result in different increments to the resistivity $\alpha = \Delta\rho/\Delta\varepsilon$; where $\alpha_S > \alpha_N$. The largest difference in α occurs for small relative elongations and can amount to $\alpha_S/\alpha_N \sim 2$-3. As ε increases, the ratio α_S/α_N decreases and is close to 1 for $\varepsilon = 30\%$. The authors attribute this difference to a higher density of point defects in the S state than in the N state.

The structure of niobium single crystals deformed by compression by 2.2% at 4.2 K in the normal and superconducting states was studied in Ref. 232 by light and electronic microscopy. It was shown that mechanical twins are formed more easily in the normal state but their development is hindered compared with the superconducting state. The density of perfect dislocations, just as the uniformity of glide, was somewhat higher in the superconducting state. The observed differences are small, but the dislocation structures which arise after deformation in the S and N states are similar.

The accumulation of crystal-lattice defects in single- and polycrystals of high purity lead in S and N states with deformation at a constant rate 2.5×10^{-4} s^{-1} was studied in Ref. 60. The experiment was performed at 4.2 K. The resistivity was determined in a longitudinal magnetic field 10 kOe. Just as for creep [231], the increment $\Delta\rho$ due to stretching in the superconducting state is larger (the difference reaches ~10% for small ε) than in the normal state. Therefore the accumulation of defects under plastic deformation in the S state is more intense than in the N state. Approximate calculations show that as a result of a superconducting transition the dislocation density changes negligibly but the concentration of point defects increases. The method of isochronous annealing of the electric resistance was used in Ref. 71 to determine more accurately the spectrum and concentration of defects arising in lead after low-temperature deformation. This established that the density of dislocations arising during deformation is independent of and the concentration of point defects is dependent on the electronic state of the superconductor. It is believed that the motion of a long dislocation segment can be a likely mechanism for the generation of excess point defects in the S state. The thermal-inertial mechanism of low-temperature deformation also attests to this.

IX. New Method for Studying the Mechanisms of Low-Temperature Plasticity

"Acting" on a crystal by means of a superconducting transition, thereby changing the electronic properties of the crystal and the dynamical behavior of dislocations, can in principle be used as a new method for investigating lowtemperature plastic deformation. Several examples of the successful use of this method are presented below.

A. Study of Dislocation Tunneling

Theoretical investigations have shown [216] that if the interaction of dislocations with Peierls barriers controls the dislocation mobility, then at temperatures ~1–2 K the motion of double inflections, which determine the dislocation mobility, by thermal activation and quantum tunneling becomes equally likely. Below these temperatures dislocation motion should become tunneling. An experimental proof of dislocation tunneling is the athermal nature of the plasticity parameters, specifically, the yield point τ_0, which in tin single crystals was observed at ~1.3 K (Figure 51) [233]. The results of a study of the influence of the electronic state on the temperature of the transition to athermality can also serve as experimental proof of the effectiveness of dislocation tunneling. It has been shown theoretically [112, 228, 234] that as dissipation (occurring at a superconducting transition) decreases, the tunneling probability should increase. Since tin is a superconductor and passes into the superconducting state at temperatures above the onset of athermality of τ_0 it was possible to determine the dependence $\tau_0(T)$ in the N and S states. The temperature of the onset of athermality was found to be sensitive to the electronic state of the sample, and in the superconducting state athermality is less pronounced but starts at a somewhat higher temperature. The latter agrees with the theoretically predicted tunneling of dislocations under dissipation conditions in metals with a Peierls mechanism [112].

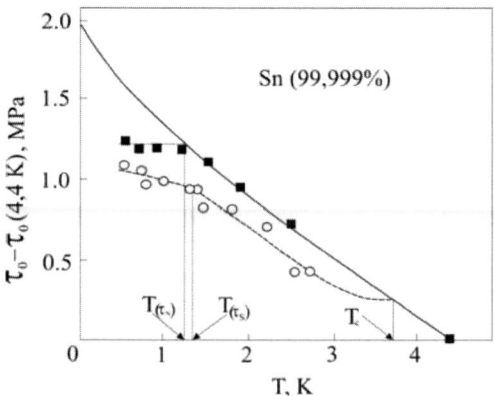

Figure 51. Temperature dependence of the yield point (τ0) of pure tin in the S (o) and N (■) states. The solid line shows the theoretical dependence τ0(T) following from the classical thermal-activation model [112]

B. STUDY OF LOW-TEMPERATURE JUMP-LIKE DEFORMATION

Experiments where controllable changes can be made in certain physical properties in the same deformable sample are important for determining the nature and specific mechanisms of low-temperature jump-like deformation. Such experiments have turned out to be experiments studying the influence of a superconducting transition on the manifestation and characteristics of jump-like deformation in aluminum [101,104,108], indium [32] lead [72, 235] aluminum alloys [154,157] lead-indium alloys [136], and tin-cadmium alloys [113]. It has been found that macroscopic jump-like deformation weakens substantially in the superconducting state, and for small deformations it vanishes (Figure 52). The observable characteristics (see the report in Ref. 109 and the review in Ref. 236) make it much more difficult to explain the low-temperature jump-like deformation on the basis of thermomechanical instability.

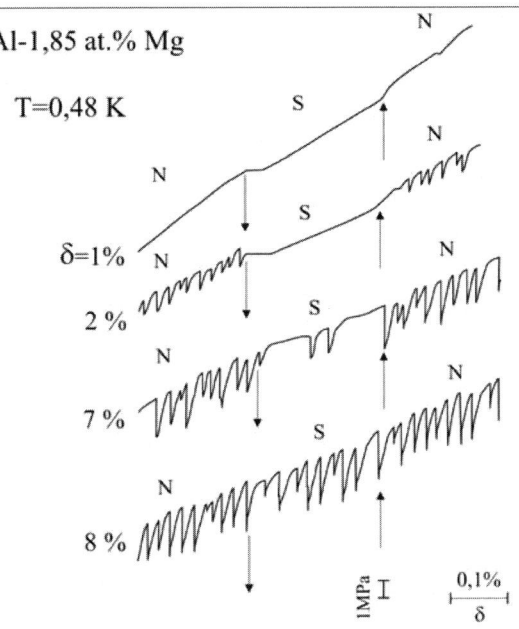

Figure 52. Sections of the tensile curves for polycrystalline Al– 1.85 at.% Mg with a change of the electronic state of the sample during deformation. T=0.48 K; $\dot{\varepsilon}$=1.1×10−4 s−1 [154]

C. STUDY OF STRESS RELAXATION

As temperature decreases, the experimentally measured degree of stress relaxation decreases, and for small degrees of deformation and sufficiently low temperatures it is not detected. A detailed analysis [237] has shown that a zero observed relaxation rate still does not mean that the true relaxation rate of plastic deformation is zero, since this rate can be too low in order to be detectable in a finite time interval with the present sensitivity of the recording apparatus. The experimental confirmation of this analysis is the sharp increase in stress relaxation at a superconducting transition, even if stress relaxation is not detected in the normal state (for example, Figure 2d). Apparently, this also occurs under creep conditions [55].

D. STUDY OF THE INTERACTION OF DISLOCATIONS AND FLUXOIDS

In principle, the interaction of dislocations with a vortex structure (boundaries of the normal phase) can also influence dislocation motion in a mixed state of a superconductor, except for the normal phase. Only calculations and observations for the motion of vortices (fluxoids) through a crystal with dislocations have been performed. The nonlinear field dependence of $\Delta\tau_{SM}$ in a mixed state indicates the presence of an appreciable interaction of a moving dislocation with vortices see (Sec. VI B 2).

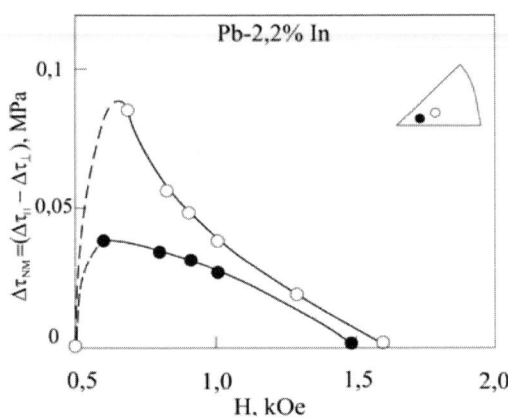

Figure 53. Additional increment in flow stress in the mixed state of single– crystalline Pb-2.2 at.% In different orientation. T=4.2 K, $\dot{\varepsilon}$=7×10−4 s−1 [174, 175]

Direct experiments performed using the influence of a superconducting transition were performed in Refs. 174 and 175 by changing the relative arrangement of the magnetic field and the glide bands, where dislocation motion occurs on deformation. An experiment consisted in deforming a crystal with a change of state of the sample from completely superconducting through mixed up to completely normal, using a magnetic field. In addition, the field had two directions with respect to the deformation axis—parallel and perpendicular. The field direction was changed in the course of the experiment by means of a mobile unit consisting of two solenoids. The objects of investigation were lead-indium alloy single crystals with indium concentration 2.2 and 12 at. %. The glide geometry of the samples was such that for a parallel field the vortices penetrated the glide surfaces at a large angle, and for a perpendicular field either they did not cross the glide plane or they crossed it at small angles. The jump of the flow stress $\Delta\tau = \tau_M - \tau_S$ was chosen as the parameter (τ_M and τ_S - are, respectively, the flow stress in the mixed and superconducting states). The quantity $\Delta\tau_{SM}=\Delta\tau_\parallel - \Delta\tau_\perp$ (Figure 53) was taken as the characteristic of the interaction; the characteristic values of this quantity are several kPa. The theoretical estimates of the largest force of the magnetoelastic interaction of a vortex with defects in a crystal give values of the order of 1 kPa. [238,239]. Considering the approximate nature of the estimates, these values are close to those observed experimentally.

E. STUDY OF THE INTERACTION OF DISLOCATIONS AND NORMALPHASE DOMAINS IN THE INTERMEDIATE STATE

The idea of the investigation examined above is close to that of Ref. 170, where the intermediate state of indium was studied. The contributions of the normal electrons and the interphase boundaries of a static intermediate state to the flow stress of single- and polycrystals of indium (99.9996%) in the temperature interval 1.7–3.4 K in an external constant magnetic field with longitudinal ($H_\parallel$) and transverse ($H_\perp$) orientation with respect to the load application axis were distinguished in this work. Comparing the relative increments of the flow stress $\Delta\tau_{IS}/\Delta\tau_{NS}$ and $\Delta\sigma_{IS}/\Delta\sigma_{NS}$ for the volume concentration of the normal phase C_N changing from 0 to 1 shows that there is no direct proportionality between $\Delta\tau_{IS}$ ($\Delta\sigma_{IS}$) and C_N, the level of hardening of a polycrystal is higher than that of a single crystal, and the stress jump for H is much larger than for H. The latter means that at a transition from a superconducting into a static intermediate state the increment to the flow stress with unchanged defect structure is determined by the drag of moving dislocations by the normal electrons and the interfaces between the normal and superconducting phases.

F. Plastic Deformation of a Composite Material

The use of a superconducting transition has been found to be effective for studying a low-temperature plastic composite material, consisting of a nonsuperconducting matrix (copper) and superconducting fibers of niobium several millimeters long with transverse dimensions 3×30 μ m [158]. The work hardening curves show three stages which are characteristic of composite materials. At the first stage both phases comprising the composite material deform elastically; at the second stage the softer phase (copper) starts to deform plastically and the harder phase (niobium) continues to deform elastically; the third stage is associated with plastic deformation of niobium. The study of the influence of a superconducting transition (switching a 12 kOe magnetic field on and off) on the flow stress showed that a jump $\Delta\sigma_{SN}$ is observed already at the second stage, where only the nonsuperconducting copper deforms. Analysis has shown that this is most likely due to the proximity effect, as a result of which the part of the copper component that is in direct contact with niobium is transferred into the superconducting state. This method was used to observe very subtle details of the deformation of the composite material.

The examples presented above do not, of course, exhaust the potential possibilities of using a superconducting transition to study plastic deformation at low temperature.

X. APPLIED ASPECTS

Among published investigations of the influence of a superconducting transition on plasticity there are works which are of interest for applications.

A. WORK HARDENING BY SUPERCONDUCTING TRANSITIONS

The various experiments presented above have shown unequivocally that at a transition of a material into the superconducting state there arises a situation which is favorable for motion of the dislocations present in the crystal and, possibly, nucleation of new dislocations. If this is so, each NS transition can change the dislocation structure and thereby harden or soften a crystal. To check this proposition experiments were performed on 99.999% pure lead polycrystals and a lead-bismuth alloy[147]. At first all samples were deformed below T_c (4.2 K) to deformation 10%. The samples were in different electronic states (the scheme is shown on the left-hand side of Figure 54): completely in a superconducting state; completely in a normal state, produced by the field of a superconducting solenoid inside which the sample is deformed ($H \approx$ 1000 Oe for lead and $\approx$ 4000 Oe for Pb-Bi alloys); with repeated changes of state. Next, all samples were unloaded, heated to temperatures above T_c (77.3 or 300 K), and deformed to failure.

If at the end of preliminary deformation the difference between the tensile curves did not exceed $\Delta\sigma_{NS}$ (~1%), then after heating to 77.3 K the yield point was found to be sensitive to the conditions of preliminary deformation. For samples which were predeformed at 4.2 K under conditions of repeated (15 times) superconducting transitions $\Delta\sigma_{NS}^{II}$ exceeded $\Delta\sigma_{S}^{II}$ - by ~54% and $\Delta\sigma_{N}^{II}$ by 40% (Figure 39, right-hand side). A somewhat smaller difference, but with the same sign, is observed in the values of the ultimate strength. Experiments with heating to 300 K qualitatively confirmed this result, though the variance was large. A similar result was also obtained for lead-bismuth alloys. Thus predeformation of a superconductor below T_c with a repeated change of plasticity as a result of repeated *NS* and *SN* transition makes it possible to produce a defect structure possessing weak recovery and high hardening, i.e. it makes it possible to obtain superconducting materials with a high yield point.

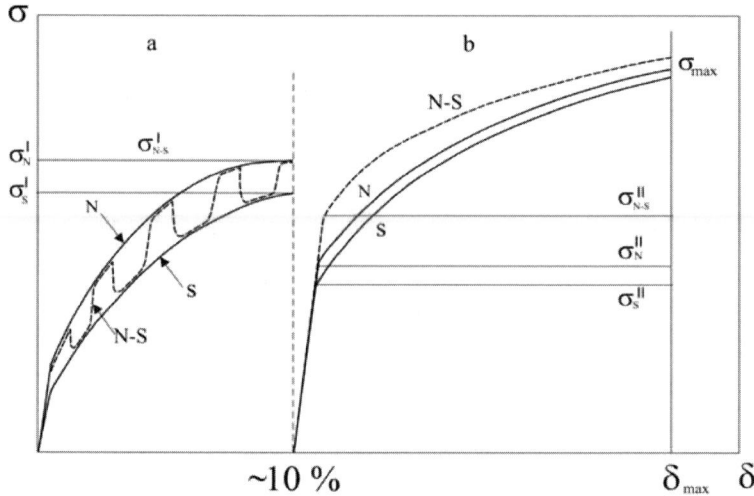

Figure 54. Scheme of an experiment on hardening of a superconductor by predeformation below Tc: deformation at 4.2 K (a); deformation after heating up to 77.3 or 300 K (b) [147]

A similar result of the hardening effect of a cyclic change of the electronic state but now without heating above T_c was also observed for single crystals of lead and leadindium alloys (up to 5 at. %) [77]. In this case the change of the normal and superconducting states was performed at a rate of two cycles per 1% relative elongation. Figure 55 shows work hardening curves $\tau(\gamma)$ for Pb–5 at. % In single crystals deformed in various regimes—in the normal state (curve *1)*, in the superconducting state (curve *2)*, and with a cyclic change of the normal and superconducting states (curve *3)*. In the latter case the plastic flow occurred under the highest flow stresses. The stress $\Delta\tau_{NS}$ was more than 85% higher than τ_N.

At all stages of deformation $\theta_{NS} > \theta_N$ and, correspondingly, $\tau_{max,NS} > \tau_{max,N}$. The uniform elongation increases by ~20%. The excess flow stress $\tau_{NS} - \tau_N$ as a result of the NS transitions grows linearly with the indium concentration. Special experiments showed that the observed hardening is due not to cycling of the stress by τ_{NS} but rather by cycling of the state. A study of the electrical resistance after loading in various regimes showed (measurements in the normal state) that the largest increment to the resistivity is observed in a sample with a cyclic change of states. This means that the density of deformation defects is highest in a sample with repeated NS transitions.

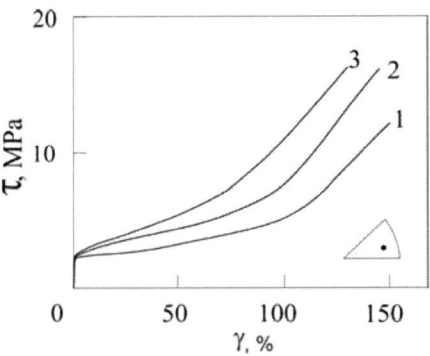

Figure 55. Tensile curves for single crystals of the alloys Pb–5 at.% In, corresponding to the normal (*1*) and superconducting (*2*) states and cyclic change of normal and superconducting states (*3*) [77]

B. WORK HARDENING IN THE SUPERCONDUCTING STATE

A new phenomenon accompanying deformation in a superconducting state—additional strain hardening—was observed in the course of systematic investigations of the influence of a superconducting transition on plasticity. The first observation was made in Ref. 101 with deformation of aluminum single crystals (99.999%) with an orientation of the stretching axis for which at 0.52 K the hardening curve consisted of three stages. Figure 56 shows the stretching curves of two identical samples, deformed below T_c at the same temperature but in different states. The yield point (left-hand inset in the figure) in the S state is lower than in the N state, which agrees with all preceding measurements. However, the work hardening curve in the superconducting state proceeds with greater hardening than in the *N* state. For this reason, as the deformation increases, the work hardening curve in the *S* state with ~5% deformation intersects the curve in the *N* state. Furthermore, the curve in the *S* state lies above the curve in the *N* state, right up to failure. A change of state in the course of deformation (right-hand inset) on both curves gives, as usual, the same result—a *SN* transition results in an increase of the flow stress and a *NS* transition results in a decrease of $\Delta\tau_{SN}$.

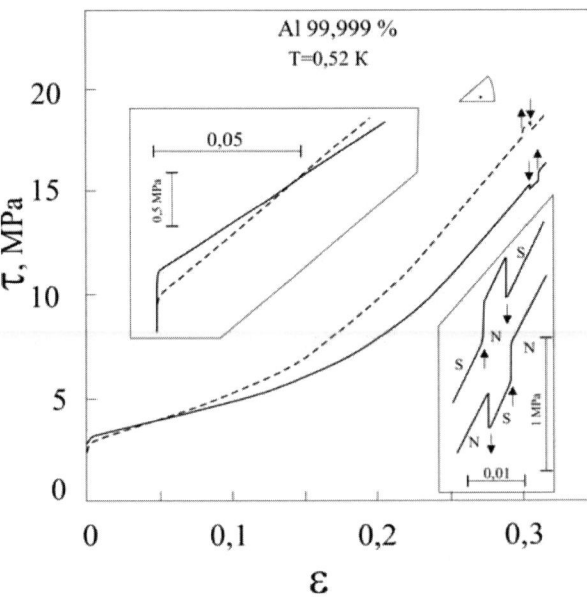

Figure 56. Tensile curves for aluminum crystals in the normal (solid line) and superconducting (dashed line) states, $T=0.52$ K; $\dot{\varepsilon} = 1\times10^{-5}$ s^{-1}. Lefthand inset: initial sections of the curves on an enlarged scale. Right-hand inset: jumps of the deforming stress at *NS* and *SN* transitions at the end of the hardening curves [101]

A similar result was obtained for the same sample when the experiment was performed under substantial deformations in each state between *NS* and *SN* transitions. The quantity θ_S/θ_N is ≈ 1.2 at the first stage and 1.05 at the second stage. Subsequently, a higher level of strain hardening was observed in the superconducting state in experiments performed on single- and polycrystals of lead (99.9996%) [131]. Resistometry shows that the reason for the difference in θ is that defects accumulate more intensely in the *S* state [77]. All the above-mentioned experiments in the *N* state were performed in an external magnetic field, which could influence the defect structure that arises. To eliminate field effects on aluminum single crystals [110] experiments were performed with stretching at temperatures slightly above T_c=1.175 K (1.3 K–N state) and below T_c (1 K–S state). For ε>0,1 the inequality $\theta_S > \theta_N$ holds with θ_S/θ_N=1.68. The results obtained mean that the softening of a superconductor, which is often called a change of plasticity at a superconducting transition, occurs only near the transition itself. At all other times additional hardening is observed in the S state. This can be explained theoretically by the facilitation of the work of the source of dislocations in the Frank–Read model in the superconducting state [240]. The result of these investigations is a new method for hardening superconductors, together with the method described in Sec. X A.

C. Fatigue in the Normal and Superconducting States

The plasticity changes observed suggested that a superconducting transition can influence the characteristics of the fatigue of a superconductor. To check this proposition experiments were performed first on lead (99.9995%) polycrystals [64,78,159]. Cyclic loading by a cantilevered bending load at 700 cycles/min was studied at 4.2 K. The constant deformation amplitude was ±0.0036 (loading rate 1.610 s−1). The experiments consisted of cyclic deformation of identical samples in the normal and superconducting states and estimating their lifetime or the number of cycles to fracture. The longevity of lead in the superconducting state is 12–60% lower than in the normal state. Oscillograph traces taken of the cycling process showed (Figure 57) that the influence of the electronic state of a superconductor was reflected in the magnitude of the plastic deformation in a cycle, especially at the start of cycling. In the first and second cycles the plastic deformation in the S state is 4–5% larger than in the N state. As the number of cycles increases, the difference in amplitudes decreases.

The situation in brittle superconducting alloys, used in commercial superconducting wires, is somewhat different [159]. The lifetime of the quasibrittle alloy 50Nb- 50Ti in the temperature range 13–293 K is essentially independent of temperature. The transition into the superconducting state (T_c=9.4–9.7 K) sharply increases the lifetime. The lifetime of the brittle alloy 50Nb–50Zr in the temperature interval 4.2–80 K is independent of temperature in the normal and superconducting states ($T_c \approx 11$ K). The difference between the two alloys under cyclic deformation is that microplasticity is observed in the quasiplastic alloy 50Nb– 50Ti, whereas it is not observed in the brittle alloy 50Nb– 50Zr.

Hence it follows that the sensitivity of fatigue to a superconducting transition is due to the existence of microplastic deformation, which a superconducting transition in- fluences.

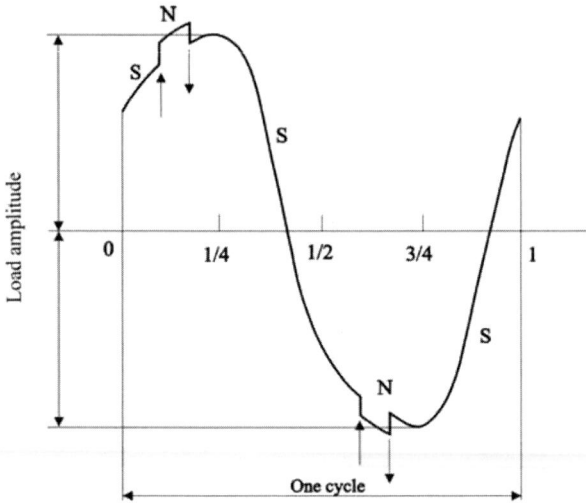

Figure 57. Oscillogram of a single cycle in experiments on fatigue of polycrystalline lead with *NS* and *SN* transitions [78]

D. INFLUENCE OF A SUPERCONDUCTING TRANSITION ON THE TRIBOLOGIC PROPERTIES OF SUPERCONDUCTORS

Since friction and wear of a material depend on its mechanical properties, the tribologic properties should be sensitive to a superconducting transition. Polycrystalline lead at 4.2 K was investigated first [241]. The rubbing unit was placed inside a superconducting solenoid. The rubbing was performed using the ring-hemispherical indentor scheme, which gives the greatest access of liquid helium to the rubbing zone, and friction was studied in the normal and superconducting states and at a superconducting transition in the course of rubbing. Under the loads investigated 1.2 and 3 kgf the friction force is larger in the *S* state than in the *N* state and correspondingly, the coefficients of friction differ, varying from 0.46 in the *N* state up to 0.53 in the *S* state. The observed changes are due to the higher plasticity of lead at a transition into the superconducting state. These changes are very clearly seen in experiments where the electronic state of a sample changed during rubbing (Figure 58). Similar results were also observed for other pairs of superconductors— niobium-niobium, vanadium-vanadium, and tantalum-tantalum [242]. The wear, as measured by weighing, of a sample under all loads was less in the superconducting state. This is due to hardening of the rubbing track due to an increase of the degree of deformability in the superconducting state. The higher strain hardening in this case could play a role (see Sec. X B).

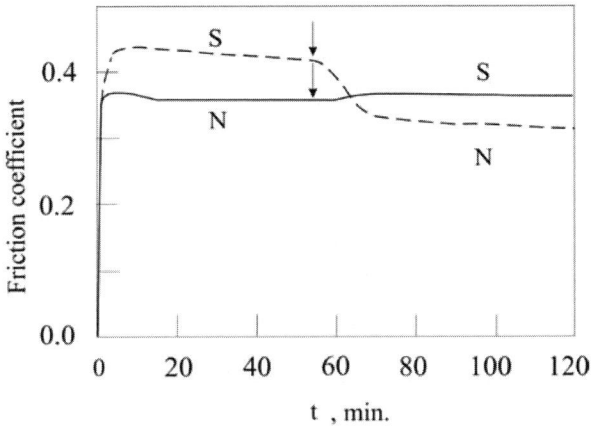

Figure 58. Influence of superconducting transitions on the friction coefficient of the pair brass-lead during rubbing in liquid helium. Velocity 0.002 m/ s. The arrows mark transitions from N into S and S into N states [239, 240]

In closing this section we present the opinion stated in Ref. 196 that the change of plasticity at a superconducting transition will give a more complete understanding of the plastic properties of superconductors as compared with nonsuperconductors. Aside from the technical and theoretical interest in superconductors, the explanation of plastic flow in superconductors is especially important for discussing plastic flow in the core of neutron stars, which are in a superfluid state.

XI. EXPERIMENTS ON HTSC

Experiments studying the mechanical properties of HTSCs were begun almost immediately after the discovery of high-temperature superconductivity. The study of the mechanical properties of single crystals and ceramic HTSCs showed (see the reviews Refs. 243 and 244) that they are low-plastic, substantially brittle materials. Practically simultaneously with the study of the mechanical properties, experiments were performed to study the possible change of the plastic properties at a transition into the superconducting state. Since the plasticity is small, the experiments on HTSCs required using highly sensitive methods. Encouraging results were obtained in works where this could be done. In Ref. 245 a laser interferometer was used to study the plasticity of the ceramics $YBa_2Cu_3O_{7-\delta}$ and $YBaSrCu_3O_{7-\delta}$; this made it possible to measure the deformation rate with high precision. The compression stress was 12 MPa, and the temperature of the experiment was 77.3 K $\leq T \leq$ 300 K (T_c of the samples was 90–95 K). The superconducting transition was accomplished by passing a current through the sample above the critical value during deformation. Deformation at 77 K was studied in detail. When current was switched on (*SN* transition), slowing (or stopping) of deformation was always observed on the interferogram. Two forms of changes are distinguished—a short-time sharp (up to tenfold) decrease of the creep rate right up to stopping of deformation for several seconds and a more prolonged but less pronounced slowing of the rate (by a factor of 1.5–2). The reverse effects were observed when the current was switched off (*NS* transition). The influence of the *NS* transition has the same sign as in superconducting metals and alloys but a much smaller magnitude, possibly, also because the measurements were performed below the yield point. In contrast to Ref. 245, in

Ref. 246 a technique used for low-temperature superconductors (for example, Refs. 26 and 27) was used to study the microplasticity of the ceramic of the orthorhombic phase of $YBa_2Cu_3O_{7-\delta}$ in the liquid nitrogen with repeated NS and NS transitions. Just as in the preceding case, the transition from the normal into the superconducting state accelerated creep, and a SN transition slowed or even stopped creep. The ratio of the strain rates in the S and N states was 2–8. In ordinary superconductors this ratio was more than two orders of magnitude. As creep diminished, the influence of the superconducting transition weakened and, ultimately, vanished. Conversely, the effect of a NS transition intensified in ordinary superconductors. In experiments where creep curves following one another in the N and S states were obtained, a sharp difference of these curves was observed. The same additional load in the S state gives rise only to elastic deformation, whereas in the S state it gives rise to well-expressed microplasticity. In the works where the sensitivity of the method could not be increased (when measuring microhardness and its temperature dependence), no effect of a superconducting transition on the microhardness of HTSC was observed to within the limits of the variance. Detailed information about investigations on HTSCs and an analysis of the results are contained in the review in Ref. 244, where it is noted that in spite of some encouraging results there is still no clear understanding of the nature of microplastic deformation of ceramics and about the influence of a superconducting transition on the plasticity of HTSCs. This is because the electronic and mechanical properties of HTSCs differ substatially from the properties of ordinary low-temperature superconductors.

XII. Conclusions

Nearly forty years have passed since the first publications reporting the observation of a strong influence of a superconducting transition on the kinetics of macroscopic plastic flow of metals-superconductors appeared. Over this time the joint efforts of experimenters and theoreticians have made established and formulated a clear physical interpretation of the basic characteristics and mechanisms of this influence. On this basis a pciture of the dislocation-electronic interaction of metals was formulated and it was shown that this interaction plays a nontrivial role in inelastic deformation processes.

It has been found that the observed changes in the characteristics of plasticity can be used as a new method for performing physical investigations of deformation at low temperatures. The observed influence of a superconducting transition on fatigue, external friction, wear, and hardening could be of great practical interest.

We are grateful to V. D. Natsik and V. P. Soldatov for critical remarks concerning the content and text of this review, which improved it.

REFERENCES

[1] Kojima H. and Suzuki T. *Phys. Rev. Lett.* 1968, Vol.21, 898.
[2] Pustovalov V. V., Startsev V. I, and Fomenko V. S., Preprint Fiz. Tekh. Inst. Nisk Temp. Akad. Nauk Ukr. SSR, Kharkov, 1968.
[3] Soldatov V. P., Startsev V. I., and Vainblat T. I., Preprint Fiz. Tekh. Inst. Nisk Temp. Akad. Nauk Ukr. SSR, Kharkov ,1969.
[4] Gindin I. Ya., Lazarev B. G., Starodubov Ya. D., and Lebedev V. P., Preprint Khar. Fiz. Tekh. Inst., No. 36, Kharkov ,1969; *Dokl. Akad. Nauk SSSR*, 1969, Vol. 188, 803 [*Sov. Phys. Dokl.* 1970, Vol.14, 1011].
[5] Didenko D. A., Pustovalov V. V., Statinova V. F., Fomenko V. S, and Dotsenko V. I. *Physics of the Condensed State*, 1970, Fiz. Tekh. Inst. Nisk Temp. Akad. Nauk Ukr. SSR, Kharkov No. 10, 26.
[6] Buck O., Alers G. A., and Tittman B. R., *Scr. Metall.*, 1970, Vol. 4, 503.
[7] Suenaga M. and Galligan J. M., *Scr. Metall.*, 1970, Vol. 4, 697.
[8] Suenaga M. and Galligan J. M., in *Physical Acoustics*, edited by Mason W. P. and Thurnston R. N., Academic Press, New York 1972, Vol. 9, 1.
[9] Pustovalov V. V., in Physical Processes of Plastic Deformation at Low Temperatures, Naukova Dumka, Kiev, 1974, 152.
[10] Kostorz G., *Phys. Status Solidi B*, 1973, Vol. 58, 9.
[11] Startsev V. I., Il'ichev V. Ya, and Pustovalov V. V., *Plasticity and Strength of Metals and Alloys at Low Temperatures*, Metallurgiya, Moscow, 1975.
[12] Suzuki T., in *Rate Processes in Plastic Deformation of Materials*, edited by J. C. M. Li and A. K. Mukherrjee, ASM, 1975, 227.
[13] Startsev V. I., *Krist. Tech.*, 1977, Vol. 14, 329.

[14] Startsev V. I., in *Dislocation in Solids*, edited by F. R. N. Nabarro, North-Holland, Amsterdam 1983, vol.6, p.145.
[15] Suzuki. T., Yoshinaga H., and Takeuchi S., *Dislocation Dynamics of Plasticity*, Springer-Verlag, New York ,1990, Mir, Moscow ,1989.
[16] D. Shoenberg, *Superconductivity*, Cambridge University Press, New York, 1952, Izd. Inostr. Lit. Moscow, 1955.
[17] Alers G. A. and Waldorf D. L., *IBM J. Res. Dev.*, 1962, vol.6, 89.
[18] Kravchenko V. Ya., *Fiz. Tverd. Tela*, Leningrad, 1966, vol. 8, 927; [Sov. Phys. Solid State, 1966, vol.8, 740].
[19] Tittman B. R. and Bömmel H. E., *Phys. Rev.*, 1966, vol.151, 187; Holstein T., *Appendix*.
[20] Berner R. and Kronmüller T., *Plastic Deformation of Single Crystals*, Mir, Moscow, 1969.
[21] Klyavin O. V. and Stepanov A. V., *Sov. Phys. Solid State*, 1959, vol.1, 241.
[22] Suzuki T. and Ishii T., *Trans. Jpn. Inst. Met.*, 1968, vol. 9, 687.
[23] Startsev V. I., Pustovalov V. V., and Fomenko V. S., *Trans. Jpn. Inst. Met.*, 1968, 9, 843.
[24] Dotsenko V. I., Landau A. I., and Pustovalov V. V., *Topical Problems of Low Temperature Plastic of Materials*, Naukova Dumka, Kiev 1987.
[25] Lebedev V. P. and Le Khak Khiep, *Fiz. Nizk. Temp.*, 1985, vol. 11, 896; [Sov. J. Low Temp. Phys., 1985, vol.11, 494].
[26] L. P. Kubin and B. Jouffrey, *Philos. Mag.*, 1971, vol.24, 437.
[27] Parasmevaran V. P. and Weertman J., *Metallorg. Trans.*, 1971, vol.2, 1233.
[28] Pustovalov V. V., Startsev V. I., and Fomenko V. S., *Fiz. Tverd. Tela*, Leningrad, 1969, vol.11, 1382; [*Sov. Phys. Solid State* 1969, 11, 1119].
[29] Pustovalov V. V., Startsev V. I., and Fomenko V. S., Preprint Fiz. Tekh. Inst. Nisk Temp. Akad. Nauk Ukr. SSR, Kharkov ,1969.
[30] Pustovalov V. V., Startsev V. I., and Fomenko V. S., *Phys. Status Solidi*, 1970, vol.37, 413 .
[31] Pustovalov V. V., in *Physics of Strain Hardening of Crystals*, Naukova Dumka, Kiev , 1972, p. 128.
[32] Kuz'menko I N., Lubenets S. V., Pustovalov V. V., and Fomenko L. S., *Fiz. Nizk. Temp.* 1983, vol. 9, 865; [*Sov. J. Low Temp. Phys.*,1983, vol.9, 450].
[33] Takeuchi S., Hashimoto T., and Maeda K., *Trans. Jpn. Inst. Met.* 1982, vol. 23, 60.
[34] Soldatov V. P., Startsev V. I., and Vainblat T. I., *Phys. Status Solidi* 1970, vol. 37, 47.
[35] Soldatov V. P., Startsev V. I., and Vainblat T. I., *J. Low Temp. Phys.* 1970, vol 2, 641.

[36] Suenaga M. and Galligan J. M., *Scr. Metall.*, 1971, vol 5, 829.
[37] Suenaga M. and Galligan J. M., *Phys. Rev. Lett.*,1971, vol 27, 721.
[38] Dotsenko V. I., Pustovalov V. V., and Fomenko V. S., *Fiz. Tverd. Tela*, Leningrad, 1972, vol 14, 201; [*Sov. Phys. Solid State*, 1972, vol 14, 161].
[39] Suenaga M.and Galligan J. M., Scr. Metall. 1971, vol 5, 63.
[40] Fomenko V. S., Startsev V. I., Pustovalov V. V., and Didenko D. A., in *Physics of the Condensed State*, Fiz. Tekh. Inst. Nisk Temp. Akad. Nauk Ukr. SSR, Kharkov 1969, No. 5, 228.
[41] Alers G. A., Buck O., and Tittman B. R., *Phys. Rev. Lett.*, 1969, vol. 23, 290.
[42] Pustovalov V. V. and Fomenko V. S., *JETP Lett.*, 1970, vol. 12, 10.
[43] Startsev V. I., Pustovalov V. V., Soldatov V. P., Fomenko V. S., and Vainblat T. I., *Proc. 2nd Intern. Conf. Str. Met. Alloys*, Am. Soc. Met., 1970, Vol. 1, 219.
[44] Bobrov V. S. and Gutmanas E. Yu., *Phys. Status Solidi B*, 1972, vol 54, 413.
[45] Fomenko. V. S, *Zh. Éksp. Teor. Fiz.* 1972, vol. 62, 2190; [Sov. Phys. JETP 1972, vol 35, 1145].
[46] Hutchison T. S. and McBride S. L., *Can. J. Phys.* 1972, vol 50, 2592.
[47] Kostorz G., *J. Low Temp. Phys.* 1973, vol. 10, 167.
[48] Kuramoto E., Iida F., Takeuchi S., and Suzuki T., *J. Phys. Soc. Jpn.* 1974, vol. 37, 280.
[49] Bobrov V. S., *Fiz. Tverd. Tela*, Leningrad, 1974, vol. 16, 3975; [*Sov. Phys. Solid State*, 1974, vol. 16, 2187].
[50] Dotsenko V. I., Pustovalov V. V., and Fomenko V. S., in *Physics of the Condensed State*, Fiz. Tekh. Inst. Nisk Temp. Akad. Nauk Ukr. SSR, 1971, No. 16, 42.
[51] Soldatov V. P., Startsev V. I., Vainblat T. I., and Danilenko L. I., Preprint Fiz. Tekh. Inst. Nisk Temp. Akad. Nauk Ukr. SSR, Kharkov 1972; *Phys. Status Solidi B*, 1972, vol. 53, 261.
[52] Soldatov V. P., Startsev V. I., and Shklyarevskaya G. I., *Fiz. Nizk. Temp.*, 1975, vol. 1, 1311; [*Sov. J. Low Temp. Phys.* 1975, vol. 1, 629].
[53] Patzak R. and Stangler F., Sitz. Mathem.-Naturwis. Klasse 5 October, Oster. Akad. Wissensch., 1972, 7.
[54] Kojima H., Moriya T., and Suzuki T., *J. Phys. Soc. Jpn*, 1975, vol. 38, 1032.
[55] Vainblat T. I., Fiz. Met. Metalloved. 1974, vol. 38, 638.
[56] Tregilgas J. H. and Galligan J. M, *Scr. Metall.* 1975, vol. 9, 199.

[57] Abraimov V. V., Soldatov V. P., and Startsev V. I., *Zh. Éksp. Teor. Fiz.* 1975, vol 68, 2185; [*Sov. Phys. JETP,* 1975, vol. 41, 1094].
[58] Hutchison T. S. and McBride S. L., *J. Low Temp. Phys.* 1976, vol. 22, 121.
[59] Hutchison T. S. and McBride S. L., *Can. J. Phys.* 1975, vol. 53, 945.
[60] Lebedev V. P.and Krylovski V. S., *Fiz. Tverd. Tela,* Leningrad, 1976, vol. 18, 3648 [Sov. Phys. Solid State, 1976, vol. 18, 2124].
[61] Rogers D. H., Hutchison T. S., and S. L. McBride, *Nuovo Cimento Soc. Ital. Fis., B*, 1976, vol. 33, 131.
[62] Patzak R. and Stangler F., *Cryst. Lattice Defects* 1974, vol. 5, 83.
[63] Esparza D. A. and Alers M., *Acta Metall.* 1977, vol.25, 1047.
[64] Verkin B. I., Grinberg N. M., Lubarski I. M., Pustovalov V. V., and Yakovenko L. F., *Acta Metall.* 1977, vol.25, 1503.
[65] Wagner D., Rainer S., and Stangler F., *J. Phys.* 1978, vol.39, 689.
[66] Lebedev V. P. and Khotkevich V. I., *Fiz. Tverd. Tela*, Leningrad, 1977, vol.19, 1295; [*Sov. Phys. Solid State* 1977, vol.19, 1343].
[67] Bobrov V. S., *Physica B*, 1981, vol.107, 721.
[68] Lebedev V. P. and Le Khak Khiep, Preprint No. 11–83, Fiz. Tekh. Inst. Nisk Temp. Akad. Nauk Ukr. SSR, Kharkov 1983.
[69] Lebedev V. P. and Le Khak Khiep, Fiz. Tverd. Tela, 1983, vol.25, 228; [Sov. Phys. Solid State 1983, vol.25, 125].
[70] Lebedev V. P. and Le Khak Khiep, Preprint No. 12–83, Fiz. Tekh. Inst. Nisk Temp. Akad. Nauk Ukr. SSR, Kharkov,1983.
[71] Krylovskiĭ V. S. and Lebedev V. P., Preprint No. 13–83, Fiz. Tekh. Inst. Nisk Temp. Akad. Nauk Ukr. SSR, Kharkov, 1983.
[72] Kuz'menko I. N. and Pustovalov V. V., *Dokl. Akad. Nauk SSSR*, 1985, vol.282, 599; [*Sov. Phys. Dokl.* 1985, vol.282, 424].
[73] Gindin I. A., Lazarev B. G., Starodubov Ya. D., and Lebedev V. P., Preprint No. 70–56, Khar. Fiz. Tekh. Inst., Kharkov ,1970.
[74] V. V. Abraimov, *Fiz. Nizk. Temp.* 1 vol. 3, 1314; *[Sov. J. Low Temp. Phys.* 1977, vol.3, 635].
[75] Abraimov V. V., Soldatov V. P., and Startsev V. I., *Metallofizika*,1980, vol.2, 66.
[76] Soldatov V. P., Startsev V. I., and Vainblat T. I., *Phys. Status Solidi B* 1971, vol.48, 381.
[77] Lebedev V. P., Krylovskiĭ V. S., and V. M. Pinto-Simoes, *Fiz. Nizk. Temp.* 1997, vol.23, 1126; [*J. Low Temp. Phys.* 1997, 23, 848].
[78] Verkin B. I., Grinberg N. M., Lyubarskiĭ I. M. , Pustovalov V. V., and Yakovenko L. F., *Fiz. Nizk. Temp.* 1976, vol. 2, 803 [*Sov. J. Low Temp. Phys.,*1976, vol. **2**, 396]

[79] Bobrov V. S. and Gutmanas É. Yu., in Abstracts of Reports at the 16th All-Union Conference on Low Temperatue Physics, Leningrad,1970, . 32.
[80] Lebedev V. P. and Le Khak Khiep, *Fiz. Met. Metalloved.* 1987, vol. 63, 1005.
[81] Lebedev V. P. and Khotkevich V. I., *JETP Lett.* 1980, vol. 32, 445.
[82] Soldatov V. P., Abraimov V. V., and Startsev V. I., *Fiz. Nizk. Temp.* 1976, vol. 2, 1244 [*Sov. J. Low Temp. Phys.* 1976, vol. 2, 606.
[83] Hutchison T. S. and McBride S. L., *Can. J. Phys.* 1972, vol. 50, 906.
[84] Gindin I. A., Lazarev B. G., Lebedev V. P., and Starodubov Ya. D., *Fiz. Met. Metalloved.* 1970, vol. 29, 862.
[85] Hutchison T. S. and Pawlowicz A. T., *Phys. Rev. Lett.* 1970, vol. 25, 1272.
[86] Hutchison T. S. and McBride S. L., *Can. J. Phys.* 1972, 50, 64.
[87] Lebedev . V. P. and Le Khak Khiep, *Fiz. Nizk. Temp.* 1985, vol. 11, 138; [*Sov. J. Low Temp. Phys.* 1985, vol. 11, 72].
[88] Gindin I.. A., B. G. Lazarev, V. P. Lebedev, and Ya. D. Starodubov, *JETP Lett.* 1976, 11, 188 .
[89] Lebedev. V. P. and Le Khak Khiep, Preprint No. 12–83, Fiz. Tekh. Inst. Nisk Temp. Akad. Nauk Ukr. SSR, Kharkov 1983.
[90] Kostorz G., *Acta Metall.* 1973, vol. 21, 813.
[91] Kostorz G., *Scr. Metall.* 4, 95 _1970_.
[92] Tabachnikova E. D., Bengus V. Z., and Efimov Yu. V., *Fiz. Met. Metalloved.* 1980, vol.50, 443.
[93] Karpov I. V, Leĭko E. B., and Nadgornyĭ É. M., *JETP Lett.* 1976, vol. 31, 428.
[94] Tregilgas J. H. and Galligan J. M., *J. Appl. Phys.* 1979, vol. 50, 399.
[95] Gindin I. A., Starodubov Ya. D., and Aksenov V. K., *Fiz. Tverd. Tela,* Leningrad, 1975, vol. 17, 1012 [*Sov. Phys. Solid State* 1975, vol. 17, 648.
[96] Bobrov V. S. and Lebedkin M. A., *JETP Lett.* 1976, vol. 38, 400.
[97] Tregilgas J. H. and Galligan J. M., *Scr. Metall.* 1977, vol. 11, 455.
[98] Takeuchi S. and Maeda K., *Tech. Rep. ISSP, Ser. A*, N 801, Tokyo, 1977.
[99] Takeuchi S.,. Maeda K, and Suzuki T., *Phys. Status Solidi A*, 1977, vol. 43, 289.
[100] Takeuchi S., Suzuki T., and Koizumi H., *J. Phys. Soc. Jpn.* 2000, vol. 69, 1727.
[101] Kuz'menko I. N., *Candidate's Dissertation in Physical and Mathematical Sciences*, Fiz. Tekh. Inst. Nisk Temp. Akad. Nauk Ukr. SSR, Kharkov, 1983.
[102] Natsik V. D., Kaufmann H.-I., Pustovalov V. V., and Kuz'menko I. N., in Abstracts of Reports at the 24th International Conference on the Member

Countries of the Council of Economic Assistance on Low Temperature Physics and Engineeering, Berlin, 1985, 178.
[103] Kaufmann H.-J., Kuzmenko I. N., Natsik V. D., and Pustovalov V. V., in *Proc. 7th Int. Conf. Str. Met. All.*, Pergamon Press, Montreal, 1985, 45.
[104] Kuzmenko I. N. and Pustovalov V. V., *Cryogenics* 1985, vol. 25, 346.
[105] Kuz'menko I. N. and Pustovalov V. V., *Fiz. Nizk. Temp.* 1979, vol. 5, 1 [*Sov. J. Low Temp. Phys.* 1979, vol. 5, 676.
[106] Kuramoto E., F. Iida, Yashimoto T., and Takeuchi S., *Scr. Metall.* 1974, vol. 8, 367.
[107] Iida F., Suzuki T., Kuramoto E., and Takeuchi S., *Acta Metall.* 1979, 27, 637.
[108] Kuz'menko I. N. and Pustovalov V. V., *Fiz. Nizk. Temp.* 1 vol. 5, 1433 [*Sov. J. Low Temp. Phys.*, 1979, vol. 5, 676].
[109] Pustovalov V. V., *Mater. Sci. Eng.*, 1997, vol. A 234–236, 157.
[110] Pustovalov V. V., Kuz'menko I. N., Isaev N. V.,. Fomenko V. S and Shumilin S. É., *Fiz. Nizk. Temp.* 2004, vol. 30, 109 [*Low Temp. Phys.* 2004, vol. 30, 82].
[111] Kirichenko G. I. and Shumilin S. É., *Fiz. Nizk. Temp.* 1986, vol. 12, 93 [Sov. *J. Low Temp. Phys.* 1986, vol. 12, 54].
[112] Natsik V. D., Kirichenko G. I., Pustovalov V. V., Soldatov V. P. and Shumilin S. É., *Fiz. Nizk. Temp.* 1996, vol. 22, 965 [*Low Temp. Phys.* 1996, vol. 22,
[113] Kirichenko G. I., Natsik V. D., Pustovalov V. V., Soldatov V. P. and Shumilin S. É., *Fiz. Nizk. Temp.*, 1997, , vol. 23, 1010 [*Low Temp. Phys.* 1997, 23, 758].
[114] Pustovalov V. V., Natsik V. D., Kirichenko G. I., Soldatov V. P., and Shumilin S. E., *Physica B*, 2000, vol. 284–288, 1253.
[115] Soldatov V. P., Natsik V. D., and Kirichenko G. I., *Fiz. Nizk. Temp.*, 2001, vol. 27, 1421 [*Low Temp. Phys.*, 2001, vol. 27, 1048].
[116] V. D. Natsik, V. P. Soldatov, G. I. Kirichenko, and L. G. Ivanchenko, *Fiz.Nizk. Temp.* 2003, vol. 29, 451 [*Low Temp. Phys.* 2003, vol. 29, 340].
[117] Abraimov V. V., Efimov Yu. V., Kozlov N. D., Savitskiĭ E. M., and Startsev V. I., *Fiz. Met. Metalloved.* 1974, vol. 38, 612.
[118] Kuz'menko I. N., Lavrent'ev F. F., Pustovalov V. V., and Salita O. P., Fiz. Nizk. Temp. 1982, vol. 8, 873 [Sov. J. Low Temp. Phys. 1982, vol.8, 440].
[119] Kalugin M. M., Natsik V. D., Soldatov V. P., and Shepel' I. A., *Fiz. Nizk. Temp.* 1993, vol. 19, 713 [*Low Temp. Phys.* 1993, vol.19, 512].

[120] Gektina I. V., Lavrent'ev F. F., Pustovalov V. V. and Shumilin S. É., *Fiz. Nizk. Temp.* 1987, vol. 13, 1025 [*Sov. J. Low Temp. Phys.*, vol. 1987, 13, 583].
[121] Hutchison T. S. and McBride S. L., *Can. J. Phys.* 1973, vol.51, 1797.
[122] Seko N., Tsubakihara H., Okada T., and Suita T., *Technol. Rep. Osaka Univ.* 1973, vol.23, 385.
[123] Vainblat T. I., Pustovalov V. V., Soldatov V. P., Startsev V. I., and Fomenko V. S. in Physics of the Condensed State, Fiz. Tekh. Inst. Nisk Temp. Akad. Nauk Ukr. SSR, Kharkov ,1970, No. 10, 99.
[124] Galligan J. M., Pang C. S., Tregilgas J. H., and Van Saun P., *J. Electron. Mater.* 1975, vol.4, 891.
[125] Galligan J. M., C. S. Pang, J. H. Tregilgas, P. Van Saun, *Philos. Mag.* 1976, vol.33, 891.
[126] Pang. C. S., Lin T. H., and Galligan J. M., *J. Appl. Phys.* 1978, vol.49, 333.
[127] Pang S. and Galligan J. M., *Phys. Rev. Lett.* 1979, vol.3, 1595.
[128] Dotsenko V. I., Pustovalov V. V., and Sirenko V. A., *Fiz. Nizk. Temp.*, 1981, vol.7, 100 [*Sov. J. Low Temp. Phys.* 1981, vol.7, 49].
[129] Sirenko V. A and Fomenko V. S., *Phys. Status Solidi A* 1982, vol.74, 454.
[130] Dotsenko V. I. and Sirenko V. A., *Fiz. Nizk. Temp.*, 1983, vol.9, 412 [*Sov. J. Low Temp. Phys.* 1983, vol.9, 209].
[131] Lebedev V. P. and Krylovskiĭ V. S., *Fiz. Tverd. Tela*, Leningrad, 1991, vol. 33, 2994 [*Phys. Solid State* 1991, vol.33, 1690].
[132] Pustovalov V. V., Sirenko V. A., and Fomenko V. S., *Fiz. Tverd. Tela*, Leningrad, 1983, vol. 25, 867 [*Sov. Phys. Solid State* 1983, vol.25, 496].
[133] Sirenko V. A., *Fiz. Nizk. Temp.* 1984, vol.10, 207 [*Sov. J. Low Temp.*, 1984, vol.10, 108].
[134] Dotsenko V. I., Pustovalov V. V., Sirenko V. A., and Fomenko V. S., *Cryst. Res. Technol.* 1984, vol.19, 1031.
[135] Vainblat T. I., Soldatov V. P., and Startsev V. I., *Fiz. Tverd. Tela*, Leningrad, 1970, vol.12, 3357.
[136] Pustovalov V. V., Fomenko V. S., and Gofman Yu. I., *Izv. Akad. Nauk SSSR, ser. fiz.*, 1973, vol.27, 2454; Preprint Fiz. Tekh. Inst. Nisk Temp. Akad. Nauk Ukr. SSR, Kharkov ,1973.
[137] Bar'yakhtar V. G., Gindin I. A., Gubin I. S., Druinski E. I., Lebedev V. P., Starodubov Ya. D., and Fal'ko I. I., *Fiz. Tverd. Tela*, Leningrad, 1973, vol.15, 2947 [*Sov. Phys. Solid State* 1973, vol.15, 1966].
[138] Dotsenko V. I., Pustovalov V. V., Fomenko V. S., and Shcherbina M. E., *Fiz. Nizk. Temp.* 1976, vol.2, 775 [*Sov. J. LowTemp. Phys.* 1976, vol.2, 382].

[139] Kostorz G., *Philos. Mag.* 1973, vol.27, 633.
[140] Pustovalov V. V. and Shepel' I. A., *Fiz. Met. Metalloved.* 1985, vol.60, 356.
[141] Soldatov V. P., Startsev V. I., and Vainblat T. I., *Phys. Status Solidi A*, 1974, vol.22, 109.
[142] Abraimov V. V., Soldatov V. P., and Startsev V. I., *JETP Lett.*, 1975, vol.21, 334.
[143] Startsev V. I., Soldatov V. P., and Abraimov V. V., *Reinstoffprobleme*, Akademie Verlag, Berlin, 1977, vol. 5, 699.
[144] V. I. Dotsenko and Sirenko V. A., *Fiz. Nizk. Temp.* 1983, vol.9, 539 [*Sov. J. Low Temp. Phys* 1983, vol.9, 275].
[145] Abraimov V. V., Kalugin M. M., and Soldatov V. P., *Cryst. Res. Technol.* 1984, vol.19, 1057.
[146] Isaev N. V., Natsik V. D., Pustovalov V. V., Shepel' I. A. and Shumilin S. É., *Fiz. Nizk. Temp.* 1992, vol.18, 911 *[Low Temp. Phys.* 1992, vol.18, 641].
[147] Didenko D. A. and Pustovalov V. V., *Fiz. Met. Metalloved.* 1981, vol.52, 400.
[148] Soldatov V. P., Abraimov V. V., and Startsev V. I., *Fiz. Nizk. Temp.* 1976, vol.2, 1244 [*Sov. J. Low Temp. Phys.* 1976, vol.2, 606].
[149] Dotsenko V. I., Fiz. Nizk. Temp. 1982, vol.8, 1078 [*Sov. J. Low Temp. Phys.* 1982, vol.8, 544].
[150] Pal'-Val' L. N., Shepel' I. A., Platkov V. Ya., and Pustovalov V. V., *Fiz. Nizk. Temp.* 1986, vol.12, 1065 [*Sov. J. Low Temp. Phys.* 1986, vol.12, 600].
[151] Abraimov V. V. and Soldatov V. P., *Fiz. Nizk. Temp.*, 1977 vol.3, 80 [*Sov. J. Low Temp. Phys.* 1977, vol.3, 39].
[152] 152 Tregilgas J. H. and. Galligan J. M, *Acta Metall.* 1976, vol.24, 1115.
[153] Isaev N. V., Fomenko V. S., and Pustovalov V. V., *Fiz. Nizk. Temp.* 1989, vol.15, 759 [*Sov. J. Low Temp. Phys.* 1989, vol.15, 429].
[154] Pustovalov V. V. and Shumilin S. É., *Fiz. Met. Metalloved.* 1986, vol.62, 171.
[155] Isaev N. V., Fomenko V. S., Shumilin S. É., Kolobnev N. I., Pustovalov V. V., and Fridlyander I. N., *Fiz. Nizk. Temp.* 1990, vol.16, 1347 [*Sov. J. Low Temp. Phys.* 1990, vol.16, 770].
[156] Pustovalov V. V., Isaev N. V., Fomenko V. S., Shumilin S. E., Kolobnev N. I., and Fridlyander I. N., *Cryogenics*, 1992, vol.32, 707.
[157] Isaev N. V., Pustovalov V. V., Fomenko V. S. and Shumilin S. É., *Fiz.Nizk. Temp.* 1994, vol.20, 832 [*Low Temp. Phys.* 1994, vol.20, 653].

[158] Dotsenko V. I., Kislyak I. F., Petrenko V. T., Startsev V. I., and Tikhonovskiĭ M. I., *Fiz. Nizk. Temp.* 1986, vol.12, 741 [*Sov. J. Low Temp. Phys.* 1986, vol.12, 420].
[159] Yakovenko L. F., Pustovalov V. V., and Grinberg N. M., *Mater. Sci. Eng.* 1983, vol.60, 109.
[160] Dotsenko V. I., Pustovalov V. V., and Sirenko V. A., *Scr. Metall.* 1981, vol.15, 857.
[161] Evetts J. E. and Wade J. M., *J. Phys. Chem. Solids* 1970, vol.31, 973.
[162] Ehrat R. R. and Kinderer L., *J. Low Temp. Phys.* 1974, vol.17, 255.
[163] Galligan J. M. and Tregilgas J. H., *Scr. Metall.* 1975, vol.9, 1321.
[164] Soldatov V. P., Startsev V. I., Vainblat T. I., and Kazarov Yu. G., Preprint Fiz. Tekh. Inst. Nisk Temp. Akad. Nauk Ukr. SSR, Kharkov ,1972; *J. Low Temp. Phys.* 1973, vol.11, 321.
[165] Lubenets S. V., Startsev V. I., and Fomenko L. S., *Fiz. Met. Metalloved.* 1981, vol. 52, 870.
[166] Gektina I. V.,. Kuz'menko I. N, Lavrent'ev F. F., and Pustovalov V. V., *Fiz. Nizk. Temp.* 1985, vol.11, 419 [*Sov. J. Low Temp. Phys.* 1985, vol.11, 227].
[167] Suzuki T., Kojima H., and Imanaka T., *Bull. Jpn. Inst. Met.* 1971, vol.10, 83.
[168] Startsev V. I., Abraimov V. V., and Soldatov V. P., *Sov. Phys. JETP* 1975, vol.69, 1834 [*Sov. Phys. JETP* 1975, vol.42, 931].
[169] Noto K., Sci. Rep. Res. Inst. Tohoku Univ. A, 1969, vol.20, 129.
[170] Lebedev V. P., *Fiz. Nizk. Temp.* 1988, vol.14, 312 [*Sov. J. Low Temp. Phys.*, 1988, vol.14, 178.
[171] Le Khak Khiep, Author's Abstract of Condidate's Dissertation in Physical and Mathematical Sciences, Kharkov University, Kharkov ,1983.
[172] Lebedev V. P. and Le Khak Khiep, *Fiz. Tverd. Tela* ,Leningrad, 1983, vol.25, 228 [*Sov. Phys. Solid State*, 1983, vol.25, 406].
[173] Pustovalov V. V., Startsev V. I., and Fomenko V. S., *Dokl. Akad. Nauk SSSR*, 1971, vol.196, 1313 [*Sov. Phys. Dokl.,* 1971, vol. 14, 1129].
[174] Kononenko V. I. and Pustovalov V. V., in Proceedings of the All-Union Conference on Low Temperature Physics, Minsk ,1976, 452.
[175] Kononenko V. I. and Pustovalov V. V., in *Proc. 5th Inter. Conf. Str. Metals and Alloys*, Pergamon Press, Aachen,1979, 505.
[176] Lebedev V. P. and Khotkevich V. I., *Fiz. Nizk. Temp.* 1979, vol.5, 89 [*Sov. J. Low Temp. Phys.* 1979, vol.5, 42].
[177] Holstein T., *Appendix Work*; Tittmann B. and Bömmel H., *Phys. Rev.* 1966, vol.151, 187.

[178] Brailsford A, *Phys. Rev.* 1969, vol.186, 959.
[179] Al'shits V. I., *Pis'ma Zh. Tekh. Fiz.*, 1974, vol.67, 2215 [*Sov. Phys. JETP* 1975, vol.40, 1099].
[180] Al'shits V. I. and Indenbom V. L., *Pis'ma Zh. Tekh. Fiz.* 1973, vol.64, 1 [*Sov. Phys. JETP* 1973, vol.37, 914].
[181] Huffman G. P. and Louat N., *Phys. Rev. Lett.* 1970, vol.24, 1085.
[182] Kaganov M. I. and Natsik V. D., *JETP Lett.* 1970, vol.11, 379.
[183] Kaganov M. I. and Natsik V. D., *Sov. Phys. JETP*, 1971, vol.60, 326 [*Sov. Phys. JETP* 1971, vol. 33, 177].
[184] Natsik V. D., *Fiz. Nizk. Temp.* 1976, vol.2, 933 [*Sov. J. Low Temp. Phys.* 1976, vol.2, 459].
[185] Bar'yakhtar V. G., Druinskiĭ E. I., and Fal'ko I. I., *Fiz. Met. Metalloved.* 1972, vol.33, 1.
[186] Kaganov M. I., Kravchenko V. Ya., and Natsik V. D., *Usp. Fiz. Nauk* 1973, vol.111, 655 [*Sov. Phys. JETP* 1974, vol.16, 878].
[187] Nadgornyi E., *Progr. Mater. Sci.* 31, edited by J. W. Christian, P. Haasen, and T. B. Massalski, Pergamon Press, 1988.
[188] Druinskiĭ E. I., *Fiz. Met. Metalloved.* 1974, vol.37, 207.
[189] Isaev N. V., Natsik V. D., Pustovalov V. V., Fomenko V. S., and Shumilin S. É., *Fiz. Nizk. Temp.* 2005, vol.31, 1177 [*Low Temp. Phys. 2005*, vol.31, 898].
[190] Natsik V. D., *Zh. Tekh. Fiz.* 1971, vol.61, 2540 [*Sov. Phys. JETP*, 1971, vol. 34, 1359].
[191] Natsik V. D., *Phys. Status Solidi A*, 1972, vol.14, 271.
[192] Natsik V. D., *Fiz. Nizk. Temp.*, 1975, vol.1, 488 [*Sov. J. Low Temp. Phys.* 1975, vol.1, 240.
[193] Indenbom V. and Estrin Y., *Phys. Status Solidi A*, 1971, 4, K37.
[194] Éstrin Yu. Z. and Indenbom V. L. in *Physical Process of Plastic Deformation at Low Temperatures*, Naukova Dumka, Kiev, 1974,
[195] Druinskiĭ E. I. and Fal'ko I. I., *Fiz. Met. Metalloved.* 1973, vol.35, 681.
[196] Druinskiĭ E. I. and Fal'ko I. I,. *Fiz. Met. Metalloved.* 1974, vol.37, 646.
[197] Bar'yakhtar V. G., Druinskiĭ E. I., and I. I. Fal'ko, *Pis'ma Zh. Tekh. Fiz.* 1974, vol.66, 1019.
[198] Natsik V. D. and Roshchupkin A. M., *Fiz. Nizk. Temp.* 1980, vol.6, 101 [*Sov. J. Low Temp. Phys.* 1980, vol.6, 49].
[199] Granato A. V., *Phys. Rev. B* 1971, vol.4, 2196.
[200] Granato A. V., *Phys. Rev. Lett.* 197, vol.127, 660.
[201] Kamada K. and Yoshizawa Y., *J. Phys. Soc. Jpn.* 1971, vol.31, 1056.

[202] Éstrin Yu., Z., *Fiz. Nizk. Temp.* 1975, vol.1, 91 [*Sov. J. Low Temp. Phys.* 1975, vol.1, 45].
[203] Schwarz R. B., Isaac R. D., and Granato A. V., *Phys. Rev. Lett.* 1977, vol.38, 554.
[204] Schwarz R. B., Isaac R. D., and A. V. Granato, *Phys. Rev. B*, 1978, vol.18, 4143.
[205] R. B. Schwarz and R. Labush, *J. Appl. Phys.*, 1978, vol.49, 5174.
[206] E. Bitzek, D. Weygaud, and P. Gumpsch, in *IUTAM Symposium on Mesoscopic Dynamics of Fracture Process and Material Strength*, edited by H. Kitagawa and Y. Shibutani, 2004, Vol.115, 45.
[207] Bitzek E. and Gumpsch P., *Mater. Sci. Eng., A*, 2005, vol.400–401, 40.
[208] Parkhomenko T. A. and Pustovalov V. V., *Phys. Status Solidi A*, 1982, vol.74, 11.
[209] Landau A. I., *Fiz. Nizk. Temp.*, 1979, vol.5, 97 [*Sov. J. Low Temp. Phys.*, 1979, vol.5, 46].
[210] Landau A.I., *Phys. Status Solidi* 1981, A vol.61, 555 [*Phys. Status Solidi A* 1981, vol.**65**, 119; A.I Landau, *ibid.* 415].
[211] Shepel' I. A., Zagoruĭko L. N., Natsik V. D., Pustovalov V. V., and Soldatov V. P., *Fiz. Nizk. Temp.* 1991, vol.17, 390 [*Sov. J. Low Temp. Phys.* 1991, vol.17, 202].
[212] Isaev N. V., Natsik V. D., Pustovalov V. V., Fomenko V. S., and Shumilin S. É., *Fiz. Nizk. Temp.* 1998, vol.27, 786 [*Low Temp. Phys.* 1998, 27, 369].
[213] Isaev N. V., Fomenko V. S., Pustovalov V. V., and Braude I. S., *Fiz. Nizk. Temp.* 2002, vol.28, 522 [*Low Temp. Phys.* 2002, vol.28, 369].
[214] Isaac R. D. and Granato A. V., *Phys. Rev. B* 1988, vol.37, 9278.
[215] Hiratani M. and Nadgorny E. M., *Acta Metall.* 2001, vol.49, 4337.
[216] Petukhov B. V. and Pokrovski V. L., *Zh. Éksp. Teor. Fiz.* 1972, vol.63, 634 [*Sov. Phys. JETP* 1973, vol.36, 336].
[217] Natsik V. D., *Fiz. Nizk. Temp.* 1979, vol.5, 400 [*Sov. Phys. JETP*, 1979, vol.5, 191].
[218] Indenbom V. L. and Éstrin Yu. Z., *JETP Lett.* 1973, vol.17, 468.
[219] Indenbom V.L. and Yu.Z. Estrin, *J. Low Temp. Phys.* 1975, vol.19, 83.
[220] Pashitskiĭ É. A. and Gabovich A. M., *Fiz. Met. Metalloved.* 1973, vol.36, 186.
[221] Chernov V. M., *Fiz. Tverd. Tela*, Leningrad, 1976, vol.18, 1194 [*Phys. Solid State* 1976, vol.18, 689].
[222] Takeuchi T., *J. Phys. Soc. Jpn.*, 1974, vol. 37, 1537.

[223] Feltham P., *Philos. Mag. A*, 1988, vol.57, 831; *J. Phys. F: Met. Phys.* 1980, vol.10, L–61.
[224] Feltham P., *Czech. J. Phys. B*, 1988, vol.38, 519.
[225] Osip'yan Yu.A., Vardanian R.A., in " *Crystal Lattice Defects & Dislocation Dynamics*",Vardanian R.A. Ed., Nova Science Publisher, Inc. Hundington, New York, US, pp.149-187.
[226] Osip'yan Yu.A., Vardanian R.A., *Zh. Éksp. Teor. Fiz.*, 1988, vol. 94, 291 *[Sov.Phys. JETP,* 1988, vol.67, 1682].
[227] Soldatov V. P., Natsik V. D., and Ivanchenko L. G., *Fiz. Nizk. Temp.*, 1996, vol.22, 1087 [*Low Temp. Phys.*, 1996, vol.22, 830].
[228] Petukhov B. V., Koizumi H., and Suzuki T., *Philos. Mag. A*,1998, vol.73, 1041.
[229] Lavrent'ev F. F., Salita O. P., and Vladimirova V. L., *Phys. Status Solidi B*, 1968, vol.29, 569.
[230] Pope D. P. and Vreeland T., Jr., *Philos. Mag.* 1969, vol.20, 1163.
[231] Gindin I. A., Lebedev V. P., and Starodubov Ya. D., *Fiz. Tverd. Tela*, Leningrad, 1972, vol.14, 2025 [*Sov. Phys. Solid State* 1972, vol.14, 1748].
[232] Gindin I. A., Lebedev V. P., and Starodubov Ya. D., *Fiz. Tverd. Tela*, Leningrad, 1975, vol.17, 1012 [*Sov. Phys. Solid State* 1975, vol.17, 648].
[233] Kirichenko G. I., Pustovalov V. V., Soldatov V. P., and Shumilin S. É., *Fiz. Nizk. Temp.* 1985, vol.11, 1206 [*Sov. J. Low Temp. Phys.*, 1985, vol.11, 611].
[234] Petukhov B. V., *Fiz. Nizk. Temp.*, 1985, vol.11, 1090 [*Sov. J. Low Temp. Phys.* 1985, vol.11, 601]; *Fiz. Nizk. Temp.* 1986, vol.12, 425.
[235] Gutmanas E. Y. and Estrin Y., *ibid.* 1985, vol.92, 137.
[236] Pustovalov V. V., Fiz. Nizk. Temp., 2000, vol.25, 515[Low Temp. Phys. 2000, vol.25, 375].; in *"Crystal Lattice Defects & Dislocation Dynamics"*, Vardanian R.A. Ed., Nova Science Publisher, Inc. Hundington, New York, US, pp.117-148.
[237] Dotsenko V. I., *Phys. Status Solidi* , 1979, vol. 93, 11.
[238] Galaĭko V. P., *JETP Lett.*, 1976, vol. 7, 230.
[239] Schneider E. and Kronmüller H., *Phys. Status Solidi B*, 1976, vol. 74, 261.
[240] Bar'yakhtar V. G., Druinskiĭ E. I., and I. I. Fal'ko, *Fiz. Tverd. Tela*, Leningrad, 1972, 14, [*Sov. Phys. Solid State* 1972, vol. 14, 1707].
[241] Verkin B. I., Guslyakov A. A., Kuleba V. I., Lyubarskiĭ I. M., and Pustovalov V. V., *Fiz. Nizk. Temp.* 1977, vol. 3, 1566 [*Sov. J. Low Temp. Phys.* 1977, vol. 3, 752].

[242] Kuleba V. I., Ostrovskaya E. L., and Pustovalov V. V., *Tribol. Int.*, 2001, vol. 34, 237.
[243] Lubenets S. V., Natsik V. D., and Fomenko L. S., *Fiz. Nizk. Temp.*, 1995, vol. 21, 475 [*Low Temp. Phys.* 1995, vol. 21, 367].
[244] Lubenets. S. V., Natsik V. D., and Fomenko L. S., *Fiz. Nizk. Temp.* 2004, vol. 30, 467 [*Low Temp. Phys.* 2004, vol. 30, 345.
[245] Peschanskaya N. N., Smirnov B. I., Shpeĭzman V. V., and Yakushev P. N., *Fiz. Tverd. Tela,* Leningrad, 1988, vol. 30, 3503 [*Sov. Phys. Solid State* 1988, vol. 30, 2014].
[246] Soldatov V. P., Natsik V. D., and Chaĭkovskaya N. M., *Fiz. Tverd. Tela,* Leningrad, 1991, vol. 33, 1777 [*Phys. Solid State,* 1991, vol. 33, 999].

INDEX

A

absorption, 4, 51, 69
absorption coefficient, 51
access, 104
accuracy, 3, 14, 18, 34, 46, 80
activation, 4, 75, 76, 80, 94
activation energy, 80
aging, 28, 31, 39
aid, 77
alloys, vi, 1, 4, 17, 18, 23, 24, 26, 27, 28, 30, 31, 33, 36, 37, 38, 39, 40, 45, 51, 53, 54, 55, 62, 64, 65, 67, 69, 72, 82, 84, 89, 94, 99, 100, 101, 103, 107
aluminum, 17, 18, 19, 20, 23, 24, 51, 52, 53, 94, 101, 102, 103
amplitude, 4, 43, 44, 45, 46, 103
Amsterdam, 112
anisotropy, 81
annealing, 9, 10, 11, 92
annihilation, 4
anomalous, 5, 36
antimony, 17, 31, 38
application, vi, 61, 78, 90, 97
Argentina, 1
Arrhenius equation, 73, 79
artificial, 31
attacks, 75
attention, 42, 43, 69
Austria, 1

B

barrier, 4, 72, 74, 80
barriers, 72, 73, 76, 80, 81, 82, 84, 93
base rate, 34
BCS theory, 51
behavior, 4, 17, 23, 27, 31, 37, 45, 57, 72, 93
bell, 26
bell-shaped, 26
bending, 87, 103
binding, 71
binding energy, 71
bismuth, 17, 38, 55, 82, 99, 100
brass, 105
Brownian motion, 78

C

cadmium, 17, 18, 23, 26, 38, 39, 55, 94
Canada, 1
ceramic, 107
ceramics, 107
chaotic, 76

chemical, 87
chemical etching, 87
classes, 76
classical, 79, 94
coherence, 80
combined effect, 61, 79
commercial, 103
components, 7
composite, 17, 98
compression, 11, 44, 45, 47, 59, 87, 89, 92, 107
computer, 72, 77
Computer simulation, 76
concentration, vi, 27, 29, 30, 31, 33, 37, 38, 39, 42, 51, 59, 60, 61, 62, 64, 65, 66, 72, 73, 74, 77, 80, 83, 92, 97, 101
concrete, 51
conduction, 4, 69, 80, 81
conductivity, 58, 81
confidence, 88
confidence interval, 88
Congress, iv
constant load, 13
constant rate, 10, 27, 30, 47, 52, 91, 92
control, 72
Cooper pair, 48, 71
Cooper pairs, 48, 71
copper, 4, 17, 98
correlation, vi, 25, 30, 49, 66
creep, 1, 10, 12, 13, 14, 15, 21, 27, 30, 33, 36, 37, 38, 43, 52, 54, 55, 63, 66, 72, 78, 80, 91, 92, 96, 107
critical value, 7, 58, 107
crystal, 11, 12, 18, 19, 20, 21, 25, 28, 33, 45, 54, 57, 60, 61, 74, 77, 80, 84, 87, 90, 91, 92, 93, 96, 97, 99
crystal lattice, 80
crystalline, 17, 96
crystals, 5, 9, 14, 18, 19, 21, 23, 27, 34, 35, 36, 37, 41, 44, 45, 46, 52, 54, 55, 60, 63, 64, 65, 72, 73, 81, 82, 84, 90, 100, 102
cycles, 100, 103
cycling, 101, 103

D

damping, 74
defects, 31, 73, 92, 97, 101, 102
deformability, 105
deformation, vi, 1, 3, 4, 7, 8, 10, 12, 13, 15, 18, 19, 21, 23, 24, 25, 26, 27, 28, 30, 31, 32, 33, 34, 35, 36, 38, 39, 41, 43, 44, 46, 47, 52, 54, 57, 66, 69, 72, 73, 74, 75, 77, 78, 80, 81, 82, 83, 84, 89, 91, 92, 94, 95, 96, 98, 99, 100, 101, 103, 107, 109
degree, 3, 14, 28, 31, 38, 39, 46, 72, 84, 95, 105
demagnetization, 28
density, 7, 46, 51, 55, 61, 77, 80, 89, 92, 101
detachment, 4, 73, 78
diffusion, 84
direct measure, 30
Discovery, 7
dislocation, 1, 3, 22, 46, 60, 64, 69, 70, 71, 73, 74, 75, 76, 77, 79, 80, 82, 84, 85, 87, 88, 90, 92, 93, 96, 99, 109
dislocation velocity, 71, 72
dislocations, vi, 1, 4, 44, 47, 51, 56, 60, 61, 62, 64, 69, 70, 71, 73, 74, 75, 76, 77, 78, 79, 80, 81, 82, 84, 87, 88, 89, 90, 91, 92, 93, 96, 97, 99, 103
displacement, 4, 90
distribution, 80
doping, 72
duration, 22

E

elastic deformation, 18, 108
elasticity, 3
electric current, 61
electric field, 60
electrical, 101
electrical resistance, 101
electron, 1, 4, 58, 64, 69, 80
electron density, 69
electron state, 58, 64

electronic, iv, vi, 1, 4, 7, 12, 18, 20, 36, 44, 46, 48, 51, 57, 61, 64, 69, 70, 71, 73, 80, 81, 84, 89, 90, 91, 92, 93, 95, 99, 100, 103, 104, 108, 109
electronic structure, 80
electrons, 4, 51, 55, 60, 61, 69, 73, 80, 81, 97
electrostatic, iv
elongation, 52, 100
energy, 51, 54, 55, 67, 72, 80, 82, 83
equilibrium, 3, 74
estimating, 81, 103
evidence, 46, 49
experimental condition, 8
expert, iv
exponential, 79

F

failure, 12, 25, 54, 99, 102
fatigue, 1, 103, 104, 109
Fermi, 69, 72, 80
Fermi surface, 69, 72, 80
fibers, 98
flow, 3, 4, 12, 14, 19, 20, 22, 23, 29, 35, 46, 47, 57, 59, 60, 61, 62, 65, 76, 81, 82, 85, 89, 96, 97, 98, 100, 102, 105, 109
fluctuations, 73
fluid, 51, 80
focusing, 2
fracture, 103
France, 1
free energy, 80
friction, 1, 71, 104, 105, 109
FS, 70, 71

G

generalization, 1, 85
generation, 71, 92
grain, 60
groups, 1, 12, 14
growth, 18, 21, 23, 25, 31, 39

H

head, 72
heat, 8, 31, 81
heat removal, 81
heating, 11, 24, 29, 30, 60, 81, 99, 100
height, 76
heterogeneity, 80
Holland, 112
hypothesis, 80, 81, 82
hysteresis, 59

I

IBM, 112
impulsive, 72
impurities, 26, 37, 54, 66
indium, 10, 12, 17, 19, 20, 21, 24, 25, 26, 27, 29, 30, 34, 35, 36, 38, 39, 40, 43, 45, 50, 51, 52, 53, 54, 55, 60, 62, 64, 72, 94, 97, 101
induction, 27, 28, 29, 30, 33, 60, 61, 62, 64
inelastic, 3, 109
inequality, 103
initial state, 60
injury, iv
instability, 95
intensity, 7, 9, 25, 57, 58, 59, 61, 62, 90
interaction, 1, 4, 60, 62, 64, 69, 81, 84, 93, 96, 97, 109
Interaction, 96, 97
interface, 61
interphase, 60, 61, 97
interpretation, 109
interval, 13, 25, 27, 34, 36, 38, 50, 51, 54, 57, 59, 67, 95, 97, 103
inversion, 65
Investigations, 17, 26
ions, 80
isotropic, 81, 89

J

January, vi
Japan, 1

K

kinetic energy, 71
kinetics, 9, 21, 73, 84, 109
kinks, 69, 72

L

laser, 107
lattice, 38, 77, 92
lead, 3, 4, 5, 9, 10, 12, 13, 14, 17, 18, 19, 21, 22, 23, 24, 25, 26, 27, 28, 29, 30, 31, 34, 35, 36, 37, 38, 39, 40, 41, 50, 51, 52, 53, 54, 55, 57, 58, 59, 60, 61, 62, 64, 66, 72, 80, 82, 91, 92, 94, 97, 99, 100, 102, 103, 104, 105
lifetime, 103
limitation, 7
linear, 25, 33, 50, 61, 71
linear dependence, 50, 71
liquid helium, 104, 105
liquid nitrogen, 108
lithium, 17, 53
localization, 43
location, 36
longevity, 103
low temperatures, 4, 18, 21, 38, 72, 73, 74, 75, 76, 79, 82, 95, 109
low-temperature, vi, 1, 43, 46, 75, 78, 81, 84, 92, 94, 98, 108

M

magnesium, 17, 23
magnetic, iv, 7, 8, 9, 10, 12, 14, 15, 18, 20, 25, 27, 28, 29, 30, 31, 32, 33, 39, 40, 51, 54, 57, 58, 59, 60, 61, 62, 63, 64, 82, 90, 92, 96, 97, 98, 102

magnetic effect, 57
magnetic field, 7, 8, 9, 12, 14, 18, 20, 25, 27, 28, 39, 51, 57, 58, 59, 60, 61, 62, 63, 64, 82, 90, 92, 96, 97, 98, 102
magnetic structure, 57, 61
magnetization, 28, 58, 59
magnetoelastic, 97
manganese, 23
mask, 45
matrix, 98
measurement, 3, 46, 67, 85
mechanical, iv, 3, 8, 80, 92, 104, 107
mechanical properties, 9, 104, 107
mechanical stress, 80
memory, vi
mercury, 17
metals, vi, 1, 5, 17, 21, 38, 50, 51, 52, 58, 69, 72, 73, 79, 82, 84, 89, 94, 107, 109
microscopy, 92
mobility, 87, 88, 90, 93
models, 3
modulus, 3, 74
molybdenum, 17, 18, 20, 84
momentum, 70
monograph, 72
morphology, 62, 66
Moscow, 111, 112
motion, 4, 22, 44, 47, 70, 72, 73, 75, 77, 78, 80, 81, 82, 85, 88, 90, 92, 93, 96, 99
multiplication, 4

N

natural, 30, 31
Nb, 24, 37, 41, 87
network, 76, 78
neutron stars, 105
New York, iii, iv, 111, 112, 122
Ni, 26, 28, 29, 66
nickel, 17, 26, 28, 54, 66
niobium, 3, 9, 11, 17, 24, 34, 44, 45, 46, 62, 63, 84, 88, 92, 98, 105
nonlinear, 4, 39, 64, 65, 96
nonlinearities, 33

normal, vi, 4, 8, 9, 11, 12, 21, 29, 34, 35, 36, 41, 44, 46, 48, 51, 59, 60, 61, 62, 63, 64, 65, 66, 69, 70, 71, 77, 81, 85, 87, 90, 91, 92, 96, 97, 99, 100, 101, 102, 103, 104, 108
normalization, 50
NS, 3, 10, 12, 13, 14, 15, 18, 19, 20, 21, 22, 27, 36, 44, 46, 47, 54, 57, 74, 77, 80, 82, 84, 85, 89, 99, 100, 101, 102, 104, 107
nucleation, 44, 45, 47, 84, 99

O

observations, vi, 1, 4, 15, 33, 44, 46, 69, 72, 75, 96
orientation, 22, 44, 46, 56, 60, 89, 96, 97, 101
orthorhombic, 108
oscillations, 4, 74, 81

P

paramagnetic, 26, 29, 54, 66
parameter, 30, 38, 45, 76, 77, 97
Pb, 22, 24, 28, 29, 30, 31, 32, 33, 35, 37, 39, 42, 45, 55, 56, 57, 58, 62, 63, 64, 65, 66, 96, 99, 100, 101
periodic, 14
perturbation, 80
phonon, 80
phonons, 4
physical mechanisms, vi
physical properties, 94
plastic, vi, 1, 3, 4, 10, 12, 17, 18, 19, 20, 21, 43, 44, 46, 49, 67, 69, 72, 73, 76, 79, 80, 81, 82, 84, 89, 92, 93, 95, 98, 100, 103, 105, 107, 109
plastic deformation, vi, 1, 3, 4, 10, 12, 18, 22, 43, 44, 46, 49, 69, 72, 73, 76, 79, 80, 81, 84, 92, 93, 95, 98, 103
plasticity, vi, 1, 3, 4, 7, 17, 18, 19, 20, 21, 29, 30, 31, 37, 43, 49, 56, 58, 62, 66, 72, 73, 78, 79, 81, 82, 83, 84, 90, 93, 99, 100, 101, 103, 104, 105, 107, 109
play, 105
point defects, 75, 80, 84, 92

polycrystalline, 22, 25, 50, 51, 57, 58, 60, 61, 62, 64, 91, 95, 104
power, 24
prediction, 85
preference, 83
preparation, iv
probability, 76, 82, 84, 94
property, iv
proportionality, 65, 97
proposition, 99, 103
pulse, 8
pulses, 60
pyramidal, 46, 89, 90

Q

quantum, 4, 73, 79, 80, 82, 83, 84, 85, 93
quantum fluctuations, 83
quasiparticle, 83
quasiparticles, 71

R

radical, 31
radius, 72
random, 78
range, 8, 18, 20, 22, 27, 30, 31, 34, 35, 36, 37, 39, 47, 50, 54, 59, 60, 62, 76, 90, 103
recovery, 100
recrystallization, 11
rectilinear, 69, 72
refrigeration, 24
relaxation, 1, 10, 14, 15, 27, 30, 38, 56, 63, 69, 95
relaxation rate, 14, 15, 95
relaxation time, 15
reliability, 9, 88
resistance, 92
resistivity, 60, 91, 92, 101
room temperature, 30, 33, 43, 87, 90
Russia, 1

S

sample, 7, 8, 9, 10, 11, 12, 14, 18, 20, 24, 25, 27, 29, 30, 33, 34, 44, 46, 47, 48, 50, 54, 57, 58, 59, 60, 61, 62, 63, 64, 82, 87, 89, 90, 91, 94, 95, 96, 99, 101, 102, 104, 107
scattering, 71
sensitivity, 3, 34, 35, 36, 37, 48, 60, 62, 95, 104, 108
series, 1, 28, 29, 34, 46, 51, 76, 88
services, iv
shear, 3, 9, 74
sign, 45, 48, 100, 107
signals, 46
silver, 17
single crystals, 3, 4, 5, 9, 11, 12, 13, 14, 18, 19, 20, 21, 22, 23, 24, 27, 31, 32, 34, 35, 36, 37, 41, 44, 45, 46, 47, 52, 54, 55, 56, 59, 60, 62, 63, 64, 65, 84, 85, 88, 89, 91, 92, 93, 97, 100, 101, 103, 107
single-crystalline, 25, 51
solid solutions, 28, 39
solutions, 30, 77
spatial, 80
spectrum, 80, 92
stages, 13, 22, 23, 65, 98, 101
strain, 12, 21, 25, 30, 35, 36, 37, 82, 83, 101, 102, 105, 108
strength, 21, 22, 25, 100
stress, vi, 1, 3, 4, 9, 10, 12, 14, 15, 18, 19, 20, 22, 23, 24, 25, 26, 27, 29, 30, 34, 35, 36, 38, 41, 43, 44, 46, 47, 48, 50, 54, 56, 57, 59, 60, 61, 62, 63, 65, 72, 79, 81, 82, 83, 84, 85, 87, 88, 89, 95, 96, 97, 98, 100, 102, 107
stress fields, 81
stretching, 9, 20, 28, 34, 44, 55, 57, 60, 61, 62, 64, 66, 82, 92, 101, 103
substances, 44
substitution, 30
superconducting, vi, 1, 3, 4, 7, 8, 9, 10, 11, 12, 14, 17, 18, 19, 21, 24, 27, 28, 29, 30, 31, 33, 34, 35, 36, 37, 40, 41, 43, 44, 45, 46, 47, 48, 49, 51, 55, 56, 57, 59, 60, 61, 62, 66, 69, 71, 73, 74, 76, 77, 80, 81, 82, 83, 84, 87, 89, 90, 91, 92, 93, 94, 95, 96, 97, 98, 99, 100, 101, 102, 103, 104, 105, 107, 109
superconducting materials, 11, 17, 57, 100
superconductivity, 31, 46, 49, 51, 56, 57, 58, 59, 107
superconductor, vi, 8, 18, 42, 51, 55, 57, 58, 59, 61, 70, 71, 73, 80, 82, 83, 84, 87, 92, 94, 96, 100, 103
superconductors, 7, 8, 9, 18, 19, 20, 25, 30, 50, 57, 59, 62, 65, 72, 81, 82, 103, 105, 108, 109
superfluid, 105
suppression, 59
switching, 7, 14, 25, 27, 57, 62, 90, 98
systematic, 34, 101
systems, 20, 39, 73, 87

T

tantalum, 17, 36, 51, 52, 84, 85, 105
T_c, 7, 8, 9, 12, 17, 18, 22, 25, 28, 31, 37, 47, 50, 51, 52, 53, 54, 57, 58, 85, 90, 99, 100, 101, 103, 107
temperature, vi, 1, 5, 7, 8, 11, 12, 21, 31, 33, 34, 36, 43, 47, 49, 50, 51, 53, 54, 55, 56, 57, 58, 60, 61, 69, 71, 73, 74, 76, 77, 79, 80, 82, 83, 85, 90, 92, 93, 95, 97, 98, 101, 103, 107
temperature dependence, 5, 31, 33, 49, 50, 51, 53, 54, 55, 56, 61, 71, 73, 74, 77, 83, 85, 108
tensile, 19, 58, 64, 95, 99
thallium, 10, 17, 21, 23, 25, 38, 55
theoretical, vi, 1, 4, 51, 56, 69, 71, 75, 77, 78, 83, 84, 88, 94, 97, 105
theory, 1, 4, 73, 76, 77, 78, 79, 80, 81, 83, 84, 85
thermal, 3, 4, 8, 58, 73, 75, 76, 77, 78, 79, 81, 82, 83, 84, 88, 90, 92, 93, 94
thermal activation, 73, 75, 76, 78, 79, 88, 93
thermal expansion, 3
thermal relaxation, 82
thermodynamic, 3, 80
threshold, 71

time, vi, 1, 3, 4, 10, 13, 14, 15, 20, 21, 27, 28, 31, 39, 46, 54, 56, 57, 60, 74, 75, 77, 82, 88, 90, 95, 107, 109
tin, 12, 17, 18, 20, 21, 23, 25, 26, 27, 28, 36, 38, 39, 41, 51, 52, 54, 66, 85, 93, 94
titanium, 17
Tokyo, 1, 4, 115
transformation, 79
transition, vi, 1, 3, 4, 7, 8, 9, 10, 11, 12, 14, 15, 17, 18, 19, 20, 21, 27, 28, 29, 30, 31, 36, 37, 43, 44, 45, 46, 47, 48, 49, 54, 56, 58, 60, 62, 66, 69, 71, 73, 74, 75, 76, 77, 80, 81, 82, 83, 84, 85, 87, 89, 92, 93, 94, 95, 96, 97, 98, 99, 100, 101, 103, 104, 105, 107, 109
transition temperature, 7, 9, 18
transitions, 10, 12, 13, 14, 18, 19, 20, 22, 25, 27, 29, 33, 40, 45, 46, 47, 51, 57, 71, 89, 100, 101, 102, 104, 105, 108
translational, 4
transmission, 61
travel, 4, 77
tunneling, 84, 85, 93
twinning, 20, 43, 44, 45, 46, 47, 48, 90
twins, 44, 45, 46, 90, 92

U

Ukraine, 1
ultrasound, 4, 69
uniform, 80, 101
uniformity, 92
USSR, 1

V

values, 4, 9, 12, 14, 17, 25, 27, 28, 31, 32, 33, 34, 36, 38, 39, 40, 42, 45, 46, 50, 57, 59, 60, 61, 62, 64, 77, 78, 81, 84, 88, 97, 100
vanadium, 3, 17, 44, 105
variance, 10, 36, 45, 46, 100, 108
variation, 8
vector, 43, 72, 74, 77
velocity, 70, 71, 73, 74, 75, 77, 90, 91
viscosity, 73, 83, 85
vortex, 31, 96, 97
vortex pinning, 31
vortices, 64, 96, 97

W

wear, 1, 104, 109
wires, 103

Y

yield, 3, 4, 10, 11, 12, 18, 21, 22, 23, 25, 27, 46, 77, 82, 85, 93, 94, 99, 101, 107

Z

zinc, 4, 17, 18, 23, 46, 47, 51, 52, 89, 90, 91
zirconium, 17